T0213126

SpringerBriefs in Electrical and Computer Engineering

More information about this series at http://www.springer.com/series/10059

Quang-Dung Ho · Daniel Tweed
Tho Le-Ngoc

Long Term Evolution
in Unlicensed Bands

Quang-Dung Ho
Department of Electronic
    and Computer Engineering
McGill University
Montreal, QC
Canada

Tho Le-Ngoc
Department of Electrical
    and Computer Engineering
McGill University
Montreal, QC
Canada

Daniel Tweed
Department of Electrical
    and Computer Engineering
McGill University
Montreal, QC
Canada

ISSN 2191-8112   ISSN 2191-8120   (electronic)
SpringerBriefs in Electrical and Computer Engineering
ISBN 978-3-319-47345-1   ISBN 978-3-319-47346-8   (eBook)
DOI 10.1007/978-3-319-47346-8

Library of Congress Control Number: 2016953660

Printed on acid-free paper

This Springer imprint is published by Springer Nature
The registered company is Springer International Publishing AG
The registered company address is: Gewerbestrasse 11, 6330 Cham, Switzerland

Preface

Global mobile traffic is expected to increase nearly tenfold by 2020 due to the increasing number of mobile-connected devices and the explosion of data-hungry mobile applications. Pushing traffic toward the network capacity quickly deteriorates the quality of service (QoS) perceived by users. Acquiring additional licensed spectrum to increase the capacity of Radio Access Networks (RANs) is increasingly expensive. Mobile operators are thus challenged by the revenue gap created by the exponential increase in mobile traffic not generating sufficient additional revenue to upgrade existing RANs. These circumstances have fostered an interest in cost-effective solutions to increase the capacity of RANs. Long-Term Evolution (LTE) in unlicensed bands (U-LTE) is among the promising solutions; however, since U-LTE is a nascent LTE technology, there are still various associated concerns and challenges needing to be addressed.

This brief first presents a comprehensive survey on U-LTE, focusing on technical issues and the impacts of this technology on neighboring networks in the shared frequency bands. Specifically, the concepts, motivations, benefits, obstacles, and coexistence requirements of U-LTE are presented. Three potential types of U-LTE—LTE-U, LAA-LTE, and MulteFire™—are explained. Next, current regulations for radio systems operating in unlicensed spectrum are reviewed. Due to the fact that technical knowledge of the medium access mechanisms employed by LTE and Wi-Fi is strongly required to understand and analyze the interactions between these two technologies when they operate in the same frequency band, high-level network architectures and technical details are presented. In particular, distinguishing features of CSMA/CA employed by Wi-Fi networks compared to standardized regulations are highlighted.

In order to capture the ongoing activities on U-LTE coexistence mechanisms, related works are surveyed with insight and observations on their limitations and concerns. This brief also presents our Network-aware Adaptive LBT mechanism (NALT) which is proposed for the LTE coexistence with Wi-Fi networks. In a nutshell, NALT monitors both channel conditions and usage activity to maximize its transmission opportunities, while maintaining fair sharing of the channel, in a way that is transparent to incumbent Wi-Fi devices. Finally, toward future working directions, in light of the survey, this brief identifies a number of open technical questions as well as related potential research issues in U-LTE.

The findings in this brief provide telecom engineers, researchers, and academic professionals with valuable knowledge and potential working or research directions when designing and developing medium access protocols for next-generation wireless access networks.

This work was partially supported by the Natural Sciences and Engineering Research Council (NSERC) through an NSERC CRD Grant with Huawei Technologies Canada.

Montreal, Canada   Quang-Dung Ho
September 2016   Daniel Tweed
                                                                    Tho Le-Ngoc

Contents

Acronyms

3G, 4G, 5G	Third, Fourth, Fifth Generation cellular networks
3GPP	3rd Generation Partnership Project, cellular network standards organization
AC	Access Category, used to distinguish priority classes in IEEE 802.11e
ACK	Acknowledgement, a positive response of reception
AP	Access Point, Wi-Fi station which acts as base station in infrastructure mode
BI	Backoff Interval, period of deferment used in IEEE 802.11/Wi-Fi
CA	Carrier Aggregation
CAP	Controlled Access Phase
CCA	Clear Channel Assessment
CFP	Contention-Free Period, in which polling is completed under the PCF
CITEL	Inter-American Telecommunication Commission
CoMP	Coordinated Multi Point
CoT	Channel Occupancy Time
CP	Contention Period, in which STAs contend with each other under the PCF
CSAT	Carrier Sensing Adaptive Transmission, a proposed LAA-LTE/Wi-Fi coexistence mechanism based on a duty cycle
CSMA/CA	Carrier Sense Multiple Access with Collission Avoidance, medium access protocol used in IEEE 802.11/Wi-Fi
CTS	Clear to Send
CW	Contention Window, range of possible back-off values a Wi-Fi station will select
D2D	Device-to-Device communication of user data

DCF	Distributed Coordination Function, a random access based MAC protocol used in IEEE 802.11 Wi-Fi
DCS	Dynamic Channel Selection
DFS	Dynamic Frequency Selection
DIFS	DCF Interframe Space, in IEEE 802.11 Wi-Fi
DL	Downlink, base station to user equipment communication
DSSS	Direct Sequence Spread Spectrum
ECCA	Extended Clear Channel Assessment
ED	Energy Detect
EDCA	Enhanced Distributed Channel Access, the LBT mechanism used in IEEE 802.11e
EDCF	Enhanced Distributed Coordination Function, a random access based MAC protocol used in IEEE 802.11e
EDGE	Enhanced Data rates for GSM Evolution
EIFS	Extended Interfrace Space, in IEEE 802.11 Wi-Fi
EIRP	Equivalent Isotropically Radiated Power
eNB	Evolved NodeB
EPC	Evolved Packet Core
ETSI	European Telecommunications Standards Institute
E-UTRA	Evolved Universal Terrestrial Radio Access
E-UTRAN	Evolved Universal Terrestrial Radio Access Network
FBE	Frame Based Equipment
FCC	Federal Communications Commission
FDD	Frequency Division Duplexing
FHSS	Frequency-Hopping Spread Spectrum
GC	Global Controller
GSM	Global System for Mobile communication
HARQ	Hybrid Automatic Repeat Request
HCCA	HCF Controlled Channel Access
HCF	Hybrid Coordination Function, a polling-based MAC protocol used in IEEE 802.11e
HetNet	Heterogeneous Network, a wireless network made up of different types of access nodes
HSPA	High Speed Packet Access
IEEE	Institute of Electrical and Electronics Engineers
IFS	Interframe Space
IoT	Internet of Things
IP	Internet Protocol
ISM	Industrial, Scientific and Medical radio bands
ITU	International Telecommunications Union
LAA/LAA-LTE	License Assisted Access LTE using LBT
LBE	Load Based Equipment
LBT	Listen-Before-Talk Medium Access Strategy
LTE	Long Term Evolution, 3GPP Releases 8 and 9
LTE-A	LTE-Advanced, 3GPP Releases 10 to current

LTE-U	LTE in Unlicensed bands using a duty-cycled coexistence mechanism
M2M	Machine-to-Machine communication of MTC data
MAC	Medium Access Control, sublayer of the Data Link Layer
MAN	Metropolitan Area Network
MIMO	Multiple-input Multiple-Output
MME	Mobility Management Entity
MTC	Machine Type Communications
NALT	Network-aware Adaptive Listen-Before-Talk, a proposed LAA-LTE/Wi-Fi coexistence mechanism
NAV	Network Allocation Vector
OFDM	Orthogonal Frequency Division Multiplexing
OFDMA	Orthogonal Frequency Division Multiple Access
PCDP	Packet Data Convergence Protocol, sublayer of the Data Link Layer in LTE
PCF	Point Coordination Function, a polling-based MAC protocol used in IEEE 802.11/Wi-Fi
PDN	Packet Data Network
P-GW	PDN Gateway
PHY	Physical Layer
PRB	Physical Resource Block
QAM	Quadrature Amplitude Modulation
QoS	Quality of Service
RAN	Radio Access Network
RAT	Radio Access Technology
RC	Regional Controller
RF	Radio Frequency
RFI	Radio Frequency Interference
RLAN	Radio Local Area Network, WLAN
RLC	Radio Link Control, sublayer of the Data Link Layer in LTE
RRC	Radio Resource Control, sublayer of the Network Layer in LTE
RSSI	Received Signal Strength Indicator
RTS	Request to Send
Rx	Receiver
SAE	System Architecture Evolution
SC-FDMA	Single Carrier Frequency Division Multiple Access
SDL	Supplemental Downlink
SDN	Software Defined Networking
S-GW	Serving Gateway
SIFS	Short Inter Frame Space
ST	Slot Time
STA	Station, a member of wireless network
TC	Traffic Classes

TDD	Time Division Duplexing
TDM	Time Division Multiplexing
TPC	Transmit Power Control
TS	Traffic Streams
Tx	Transmitter
TXOP	Transmission Opportunity
UE	User Equipment
UL	Uplink, user equipment to base station communication
U-LTE	LTE in Unlicensed bands, catch-all phrase to cover all methods of using LTE in unlicensed frequency bands
UMTS	Universal Mobile Telecommunications System
VoIP	Voice over IP
VoLTE	Voice over LTE
WLAN	Wireless Local Area Network

Chapter 1
Introduction

The increasing penetration of mobile-connected devices and the emergence of numerous data-hungry mobile applications have created a wide range of business opportunities for mobile network operators and service providers. Meanwhile, this growth is placing a greater pressure on the capacity that the Radio Access Networks (RANs) have to provide. This chapter begins with a review of technical challenges that motivate developments of cost-effective solutions to improve the capacity of RANs. The technology that utilizes the unlicensed frequency bands in Long-Term Evolution (LTE), namely U-LTE, is singled out as the most promising solution. Benefits and obstacles of this technology are then presented. Next, three typical forms of U-LTE including LTE-U, LAA-LTE, and MulteFire™ are explained. Concerns on the interactions between U-LTE and Wi-Fi in radio channel access and their coexistence are discussed in detail. Finally, key requirements for the coexistence of U-LTE and Wi-Fi are summarized.

1.1 Motivations and Concepts of U-LTE

Global mobile traffic is expected to increase nearly tenfold between 2014 and 2020 due to the increasing number of mobile-connected devices and the explosion of data-hungry mobile applications [1]. The expected growth in traffic is shown in detail in Fig. 1.1, where the 3.7 exabytes per month seen in 2015 are forecast to grown at a compound annual growth rate of 53 %, reaching 30.6 exabytes per month by 2020. Pushing traffic toward the network capacity quickly deteriorates the Quality of Service (QoS) perceived by the users. As a result, increasing the capacity of their Radio Access Network (RAN) is one of the top-priority action plans of mobile service providers. Purchasing additional licensed spectrum is a straightforward solution to this but radio spectrum is very much limited and increasingly expensive. Furthermore, mobile operators are at the same time challenged by the "revenue gap"; i.e., the exponential increase in mobile traffic does not generate sufficient additional revenues which would be required for upgrading their RANs. These circumstances have

© The Author(s) 2017
Q.-D. Ho et al., *Long Term Evolution in Unlicensed Bands*,
SpringerBriefs in Electrical and Computer Engineering,
DOI 10.1007/978-3-319-47346-8_1

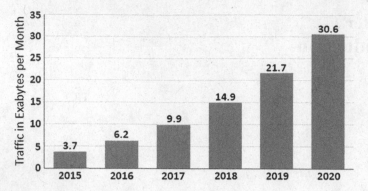

Fig. 1.1 Global mobile data traffic from 2015–2020 [1]

fostered interest in cost-effective solutions to increase RAN capacity. Mobile data offloading and Long Term Evolution (LTE) in unlicensed bands (U-LTE) are among the most promising solutions.

Mobile data offloading is the use of a complementary wireless technology to transport data originally flowing through the cellular mobile network. Wi-Fi off-loading and device-to-device (D2D) communications are the two main data offload techniques. Rules determining when and how the mobile offloading actions are triggered are set by either mobile subscribers or network operators. For the subscribers, data offloading helps them to exploit the availability of higher bandwidth data service at lower costs. For the operators, the most obvious benefit of this kind of approach is the mitigation of cellular mobile network load and thus congestion. Besides, shifting data to a complementary wireless technology leads to a number of other improvements including an overall increase in network throughput, a reduction of content delivery time, the extension of network coverage and increase of network availability, and better energy efficiency. Unfortunately, these benefits come with a number of challenges related to infrastructure coordination, network/technology handovers, service continuity, pricing, business models, and lack of existing standards.

Recently, U-LTE has appeared as the most promising approach to enhance RAN capacity and address the revenue gap in mobile networks. The original idea of LTE-U is fairly straightforward: By definition, it is an LTE technology that puts cellular signals into the unlicensed spectrum with the support of existing LTE features including supplemental downlink (SDL, proposed in LTE Release 9 and later) and carrier aggregation (CA, proposed in LTE Release 10 and later). As mentioned, mobile operators are facing a great pressure on capacity and cost. If LTE can exploit the unlicensed band (which IEEE 802.11/Wi-Fi and other radio systems are using), then it will obtain a considerable additional capacity at a minimal cost. U-LTE can be used to boost downlink (DL) or both uplink (UL) and DL of the LTE networks, as illustrated in Fig. 1.2.

Historically, U-LTE was originally proposed and officially announced by Qualcomm in 2013 [2]. Currently, it focuses on 500 MHz of spectrum available in the

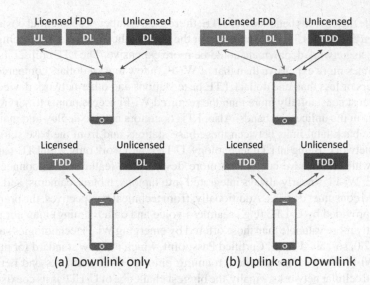

Fig. 1.2 Use cases of U-LTE

5 GHz band. Specifically, according to the proposal from Qualcomm, U-LTE is to use the U-NII-3 part of the 5 GHz band, which has highest allowed Equivalent Isotropically Radiated Power (EIRP). While in the 2.4 GHz band, regulatory bodies limit EIRP to 100 mW (in Europe) or 200 mW (in USA), and the U-NII-3 enjoys the right to go as high as 1000 mW outdoors.

1.2 Benefits and Obstacles of U-LTE

U-LTE is expected to offer numerous benefits to mobile network operators, service providers, and consumers. Free access to the unlicensed spectrum provides additional capacity to the network at a minimal cost compared to purchasing licensed spectrum allocations. Therefore, U-LTE appears to be a very inexpensive way to meet the future traffic growth. U-LTE will give operators the option to make use of unlicensed spectrum within a unified network, offering potential operational cost savings, improving spectral efficiency, and providing a better user experience. Compared to the Wi-Fi offloading technology, U-LTE has the potential to offer significantly better coverage and higher spectral efficiency while allowing seamless flow of data across licensed and unlicensed channels in a single core network. U-LTE could also take advantage of the robust security features which are already in place in LTE networks, rather than relying on external or complementary networks. Finally, Wi-Fi offloading leads to less traffic on mobile networks and thus may result in revenue losses in data services. Since U-LTE could be managed through a single core network, it could provide an incremental ability for mobile service providers to directly bill for data usage.

Despite the obvious benefits of U-LTE, there are a number of substantial obstacles. First, even though U-LTE is not charged for the use of unlicensed spectrum, compared to Wi-Fi, its network deployment could be more expensive. The LTE chipset itself is several times more expensive than that of Wi-Fi (a few tens of dollars compared to a few dollars or less than one dollar). LTE base stations and other network devices are likely to cost substantially more than the required Wi-Fi access points to service the same area in the unlicensed bands. Also, LTE operators need to deploy and maintain expensive back-haul links between these base stations and from the base station to the core network. Being an LTE technology, U-LTE will work only with LTE-capable devices while there have been many more devices that feature Wi-Fi connectivity than LTE. Wi-Fi is nearly always integrated into laptops, tablets, cameras, and other connected consumer devices. Additionally, from technical perspectives, the premium features provided by U-LTE (e.g., seamless voice and data roaming) may not prove sufficiently more valuable than those offered by emerging Wi-Fi technologies such as Hotspot 2.0, so-called Wi-Fi Certified Passpoint, which is a new standard for public-access Wi-Fi that enables seamless roaming among Wi-Fi networks and between Wi-Fi and cellular networks. Finally, the biggest challenge of U-LTE is its coexistence with other radio networks operating in the same frequency bands.

1.3   Three Types of U-LTE

There are three different flavors of U-LTE currently under development: LTE unlicensed (LTE-U), licensed-assisted access LTE (LAA-LTE), and MulteFire™. The first two flavors require "anchoring licensed spectrum"; i.e., they operate primarily in licensed spectrum and opportunistically exploit access to the unlicensed spectrum for an additional bandwidth boost. Devices are still anchored in licensed spectrum for LTE management/control signaling and high QoS data while using the unlicensed spectrum for only best-effort or delay-tolerant data. The third flavor is developed by Qualcomm and requires no licensed spectrum at all; therefore, it is often referred to as standalone U-LTE. MulteFire is designed for indoor use and deployments by enterprises, cable companies, and other service providers without ownership of expensive bandwidth licenses. However, at the present time, there are very few technical details available about MulteFire.

1.3.1   LTE-U

LTE-U is the simplest form of U-LTE and requires only minor modifications in LTE protocol stack. Therefore, it can quickly facilitate prestandard equipment manufacturing and deployment. LTE-U first attempts to select a clear channel to access. If no clear channel is found, then it will employ carrier-sense adaptive transmission (CSAT), which is a time division multiplex (TDM) coexistence mechanism based

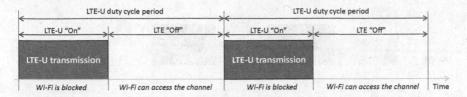

Fig. 1.3 Duty-cycling mechanism employed by LTE-U

on medium sensing. The CSAT mechanism is depicted in Fig. 1.3. CSAT employs duty-cycling, which enforces the TDM cycle on other users of the channel, instead of a LBT mechanism. In particular, CSAT defines a time cycle where the base station transmits in a fraction of the cycle and then gates off in the remaining duration. Compared to LBT or CSMA, the base station senses the medium for a longer duration (around tens of milliseconds to 200 ms) and according to the observed medium activities, the algorithm gates off LTE transmissions proportionally. The duty cycle of transmission versus gating off is dictated by the sensed medium activity of neighboring RANs. The TDM cycle can be set to a few tens or hundreds of millisecond, which can effectively accommodate the activation/de-activation procedures while controlling the data transmission delay. An important observation in Fig. 1.3 is that during the LTE "on" period, Wi-Fi is blocked by LTE-U transmissions. During the LTE "off" period, Wi-Fi will detect that the channel is free and can schedule its transmissions following its CSMA-CA protocol.

LTE-U is only applicable in areas where there are no strict LBT requirements for operations in unlicensed bands (e.g., USA, Korea, and China). It is a nonstandard version of U-LTE, being developed outside of the 3GPP standards process. LTE-U is supported by the LTE-U Forum formed in 2014 by Verizon in cooperation with Alcatel-Lucent, Ericsson, Qualcomm Technologies Inc. (a subsidiary of Qualcomm Incorporated), and Samsung.

1.3.2   LAA-LTE

In many regions (e.g., Europe, Japan, and India), there exist regulations for accessing the unlicensed spectrum that requires equipment to periodically check for the presence of other occupants in the channel, so-called LBT. LAA-LTE is designed not only for use in such areas, but also for global use. It requires a number of modifications so that LTE transmissions can meet regulatory requirements in LBT regions. Similar to LTE-U, LAA-LTE first tries to choose the cleanest channel available, based on Wi-Fi and LTE measurements. In the event that no clean channel is available, a LBT algorithm is used to compete for the medium with other RANs. For LBT, different mechanisms for frame-based equipment (FBE) and load-based equipment (LBE)

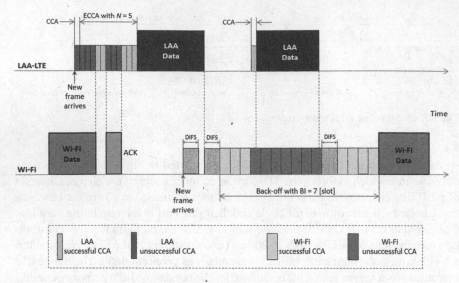

Fig. 1.4 CCA and ECCA mechanisms employed by LAA-LTE

have been specified in [3]. Details of these types of the devices and the applicable mechanisms are presented in Chap. 2.

Assuming that LBE LBT is employed for LAA-LTE, before transmission, a Clear Channel Assessment (CCA) using an energy detect (ED) threshold is performed. If the channel is clear during a CCA slot (20 μs or longer), transmission is started immediately; otherwise, an extended CCA (ECCA) is performed. If the channel is clear during N CCA slots transmission is started immediately, where N is a random integer uniformly distributed from 1 to q, and $q \in \{4, 5, \ldots, 32\}$ is a predetermined constant. The total time to occupy the channel without CCA is limited to $(13/32)q$ ms (e.g., 13 ms when q is 32). Two simplified scenarios with LAA-LTE (employing LBE-based LBT) and Wi-Fi systems operating in the same channel are illustrated in Fig. 1.4. In the first scenario, the LAA-LTE system, upon having data frames to send, performs CCA and then ECCA (with $N = 7$) since there is an ongoing Wi-Fi transmission. The ECCA procedure is frozen and then resumed when another Wi-Fi transmission takes place and then completes, respectively. The LAA-LTE system finally transmits its frames once its ECCA counter reaches zero. In the second scenario, the Wi-Fi system, upon having data frames to send, performs CCA and then backoff procedure (with $BI_{slots} = 7$) since there is an ongoing LAA-LTE transmission. The backoff procedure is frozen and resumed when another LAA-LTE transmission takes place. The Wi-Fi system finally transmits its frames once its backoff counter reaches zero.

Since LTE was originally designed for licensed spectrum and a centralized management (i.e., network-controlled) model, it is generally an "always-on" technology. As a result, adapting to LBT is a marked change for the LTE protocol. Compared to LTE-U, which is downlink-only in the unlicensed bands, LAA-LTE may be used to support bidirectional traffic in unlicensed bands. LAA-LTE is currently actively supported by 3GPP and is included in 3GPP LTE Release 13. T-Mobile USA and Verizon Wireless have indicated their interests in deploying prestandard LAA-LTE systems for evaluations and commercial services in 2016.

1.3.3 MulteFire

At this time, there are very few technical details available about MulteFire. It is unknown which MAC protocols or coexistence mechanisms are employed in this type of U-LTE. Also, since licensed frequency is not used for LTE network management and control signaling, as opposed to the conventional LTE and the other two variants of U-LTE (i.e., LTE-U and LAA-LTE) that are license anchored, MulteFire may lose all advantages of native LTE technologies. It is expected that MulteFire will be less efficient than LTE-U and LAA-LTE and therefore, its achievable performance/efficiency may be just marginally better than that of Wi-Fi. Then the question on the applicability of MulteFire needs to be answered.

1.4 Coexistence of U-LTE and Wi-Fi

It is well known that multiple radio communications technologies operating in a common frequency band will negatively affect each other if respectful coexistence mechanisms are not employed. As a result, despite the fact that U-LTE can offer various benefits, its coexistence with Wi-Fi and other radio systems that operate in the 5 GHz frequency band is the biggest concern. In specific, it is believed that U-LTE may considerably interfere with Wi-Fi systems and/or grasp more radio resources when they operate in the same frequency band due to the following:

- *First*, LTE was originally designed to work in its own licensed frequency band rather than to coexist with other radio access technologies in a shared band. LTE employs orthogonal frequency division multiple access (OFDMA) and transmits almost continuously without requiring or implementing any mechanism for spectrum sharing. Wi-Fi, on the other hand, employs a Listen-Before-Talk (LBT) medium access strategy. In fact, the Wi-Fi LBT protocol includes a few key additional features that go beyond the LBT requirements specified by European Telecommunications Standards Institute (ETSI) [3]. As a result, U-LTE might overwhelm Wi-Fi neighbors with its aggressive transmission profile, if no relevant coexistence measure is implemented.
- *Second*, the typical duration of transmission for LTE and Wi-Fi is not the same. LTE, due to its basic protocol design and scheduled nature, generally transmits long frames (i.e., multiple ms), whereas a large percentage of Wi-Fi frames are sub-millisecond in duration. For this reason, equitable access to the medium, eval-

uated in terms of how often a technology is able to start a transmission, does not necessarily translate into equitable airtime.

• *Third*, a license-anchored system (LTE-U or LAA-LTE) operates simultaneously in licensed and unlicensed bands. Thus such systems can dynamically move traffic between the bands on a granular basis (e.g., per-user and per-flow). As a result, such a system is inherently less sensitive to collisions and congestion in the unlicensed bands than a system operating solely in the unlicensed spectrum bands. This reduced sensitivity to the issues which arise from coexistence may reduce the incentive for a license-anchored system to develop effective coexistence mechanisms.

In fact, U-LTE is still a nascent LTE technology with many technical details to be determined. Proponents of U-LTE include Qualcomm, Ericsson, Alcatel-Lucent, Huawei, LTE-U Forum, 3rd Generation Partnership Project (3GPP), Verizon Wireless, and T-Mobile US. At the same time, CableLabs, Google, Wi-Fi Alliance, the Institute of Electrical and Electronics Engineers Standards Association (IEEE-SA), and many Wi-Fi-interested companies are participating in and following closely the development of U-LTE technology. These organizations have been expressing their concerns and the critical need for strong coexistence between U-LTE and Wi-Fi to ensure responsible and fair use of the unlicensed spectrum. As a result, various studies on the coexistence of U-LTE and Wi-Fi have been carried out by both industry and academia. Besides, a number of reports and comments related to this concern have been filed with the Federal Communications Commission (FCC).

1.5   Requirements of U-LTE Coexistence Mechanisms

Even though the unlicensed bands may be used by anyone, there is a series of government guidelines and regulations which must be followed. These guidelines and regulations aim to ensure that different radio systems operating in the same frequency bands are good neighbors to each other. In particular, for coexistence with Wi-Fi, at minimum U-LTE must satisfy local regulations such as the maximum transmission power in specific bands and the avoidance of bands dedicated to protected services. Furthermore, a U-LTE system should not cause any higher interference to a neighboring Wi-Fi system than a typical Wi-Fi system operating on the same channel. In other words, the impact of a U-LTE device to Wi-Fi devices (in terms of collision rate and probability of successful channel access) should be similar to that caused by a typical Wi-Fi device. These requirements ask for inclusion of a number of new features in LTE. For example, U-LTE should select a carrier which is least occupied in the area and should dynamically change operating frequency to avoid conflict over protected systems, such as radar. It should also apply LBT or Clear Channel Assessment (CCA) techniques to check that a channel is free before making a transmission. Exactly how these decisions are made will be key aspects of U-LTE system designs.

1.6   Structure of this Brief

This Springer Brief is divided into seven chapters. This introductory chapter has provided an overview of the emerging U-LTE technology and its accompanying motivations, concepts, benefits, and challenges. The technical concerns surrounding the coexistence of this new LTE technology and the existing Wi-Fi networks operating in the 5 GHz unlicensed frequency bands have also been described. The remainder of this brief is organized as follows:

For background knowledge, Chap. 2 provides an overview on radio spectrum and related management/allocation concerns in the 5 GHz unlicensed frequency band. It also summarizes a number of the key requirements and regulations specified by the European Telecommunications Standards Institute (ETSI) and the Federal Communications Commission (FCC) applying to radio channels, operating channel selection, transmission power, and channel access rules. These technical details are the baselines to be followed when designing medium access control (MAC) protocols for U-LTE and any other technologies which are designed for shared bands.

Chapter 3 presents a high-level overview of LTE-Advanced (LTE-A) networks and associated technologies to form a basis for discussion of the coexistence issues that exist for unlicensed LTE and Wi-Fi. Understanding the underlying architecture and protocols employed in LTE-A will provide readers a comparative framework to grasp how, and at what levels, LTE and Wi-Fi networks may interact and interfere with each other, and form a greater understanding of the challenges to be addressed in designing coexistence mechanisms.

Furthering the basis for framing the potential problems with U-LTE/Wi-Fi coexistence, Chap. 4 focuses on Wi-Fi technology, beginning with an overview of the evolution of IEEE 802.11/Wi-Fi. Both existing Wi-Fi generations and the next generation of IEEE 802.11/Wi-Fi currently under development are presented. The majority of this chapter provides the underlying ideas and detailed mechanisms of the CSMA/CA MAC protocol used in Wi-Fi. Important observations on how CSMA/CA senses and occupies the radio medium when the LTE network is operating in vicinity are highlighted.

Chapter 5 provides a current literature survey to present a big picture of the research activities related to the coexistence of U-LTE and Wi-Fi technologies. The following questions are addressed in this chapter: What issues arise from simultaneous operation of LTE and Wi-Fi in the same spectrum bands? Which technology is affected the most? Which factors determine the impacts of U-LTE to Wi-Fi? Finally, the chapter identifies the strengths and weaknesses of existing solutions and suggests potential strategies to improve the performance of these two technologies.

In Chap. 6, a Network-aware Adaptive Listen-Before-Talk mechanism (NALT) proposed for U-LTE is presented. To promote effective coexistence in terms of channel occupancy time, NALT passively monitors both channel conditions and usage activity to maximize its own transmission opportunities while respecting fair sharing of the channel, in a way that is transparent to incumbent Wi-Fi devices. Simulation results are presented, demonstrating the effectiveness of NALT in providing proportional fair sharing among contending LAA-LTE and Wi-Fi devices.

Finally, Chap. 7 ends this Springer Brief by addressing a number of research issues and associated potential research directions. Potential solutions to the issues identified are also discussed. Most of the solutions suggest the cooperation of LTE and Wi-Fi so that they could have a better understanding of each other when operating in the same area using the same radio frequency band. This understanding is used to have more vigilant actions that help to avoid aggressive channel access that could corrupt ongoing transmissions and to design relevant protocols for fair spectrum sharing.

References

1. "Cisco visual networking index: Global mobile data traffic forecast update 2014-2019," white paper, Cisco, 2015.
2. "Extending LTE advanced to unlicensed spectrum," white paper, Qualcomm Inc., Dec. 2013.
3. *ETSI EN 301 893 V1.7.2 (2014-07): Broadband radio access networks (BRAN); 5 GHz high performance RLAN; Harmonized EN covering the essential requirements of article 3.2 of the R&TTE Directive*, European Telecommunications Standards Institute Std., 2014.

Chapter 2
Requirements and Regulations in the 5 GHz Unlicensed Spectrum

Licenses and fees are not required for operators to use the 5 GHz unlicensed spectrum. However, in order to avoid interference and to ensure a fair use of this resource, numerous requirements and regulations are imposed by national and international organizations. When operating in this band, the emerging U-LTE technology needs to adhere to these regulations, as any other existing technologies, especially IEEE 802.11/Wi-Fi, would. This chapter provides an overview on the radio spectrum resources in these bands and the related management and alloca- tion concerns. It then summarizes a number of the key requirements and regulations specified by the European Telecommunications Standards Institute (ETSI) and the Federal Communications Commission (FCC) on radio channels, operating channel selection, transmission power, and channel access rules. These technical details are the baselines to be followed when designing medium access control protocols for U-LTE and any other technologies operating in the 5 GHz unlicensed radio band.

2.1 Radio Spectrum Management: An Overview

The radio frequency (RF) spectrum is the part of the electromagnetic spectrum from 3 Hz to 3000 GHz (3 THz). Radio waves in this frequency range are widely used in modern technologies, especially in telecommunications. The radio spectrum is divided into different chunks or bands, each of which can be used by one or multiple technologies. Radio Frequency Interference (RFI) can disrupt and disturb the nor- mal functioning of devices, and thus, it is always important to avoid or keep the RFI within acceptable levels. For this, the generation and transmission of radio waves is strictly regulated by national laws, coordinated by international organizations, e.g., Federal Communications Commission (FCC), Inter-American Telecommunication Commission (CITEL), International Telecommunication Union (ITU), and the Euro- pean Telecommunications Standards Institute (ETSI).

© The Author(s) 2017
Q.-D. Ho et al., *Long Term Evolution in Unlicensed Bands*,
SpringerBriefs in Electrical and Computer Engineering,
DOI 10.1007/978-3-319-47346-8_2

Most countries consider RF spectrum to be a national resource. The process of regulating the use of this resource is spectrum management or allocation. Spectrum allocation varies by country and/or regulatory domain. In the USA, for example, the FCC regulates interstate communications by radio, television, wire, satellite, and cable in all states and territories. From a management perspective, radio bands are categorized into *licensed* and *unlicensed*. Licensing is a way of ensuring that wireless operators do not interfere with each other by giving each of them the exclusive use of one or more bands in given geographical areas, over a set period of time. Licensed bands are mainly sold or assigned to operators through a spectrum auction process. These licensed spectrum allocations are primarily used by the television broadcasting, commercial radio, and cellular voice and data industries. Operating in their own licensed bands, operators can avoid RFI from other users and thus guarantee the quality of services they deliver to their subscribers. However, licensing would be very impractical for certain use cases, such as communications between cordless handsets and base units. Instead, such wireless technologies transmit radio signals in unlicensed frequency bands—usually the Industrial, Scientific and Medical (ISM) band defined by the ITU radio regulations and allocated in most countries for free use by anyone without any licenses and fees. Unlicensed bands enable numerous technologies and products, e.g., Wi-Fi, Bluetooth, and many other low-power short-range communications technologies. These bands are open sandboxes where users can operate without the high barriers to entry, such as cost, that are seen in the licensed spectrum bands. The availability of unlicensed bands provides a platform for innovation, a greenfield space for technology startups and entrepreneurs, as well as established companies. Internet of Things (IoT)—the development and deployment of networking technologies that provide connectivity to everyday objects for many innovative applications—is fundamentally enabled by unlicensed spectrum.

Today, most people are within a few meters of consumer products (e.g., microwave ovens, Wi-Fi, and Bluetooth) that use the unlicensed bands. In other words, there is a great chance for RFI in these bands. As a result, even though no permission is required for the use of unlicensed bands, manufacturers and users must comply with numerous rules and regulations (related to transmission power, transmission time, etc.) in order to minimize RFI to others as well as to ensure a fair sharing of the radio resource in these bands. IEEE 802.11/Wi-Fi is the most successful and popular technology operating in unlicensed spectrum. Wi-Fi manufacturers need to obtain compliance certifications from Wi-Fi Alliance whose certification program is designed following rules imposed by radio spectrum management organizations/authorities such as ETSI and FCC.

The two most widely used unlicensed bands are 2.4 and 5 GHz. These two bands have their own advantages and disadvantages in various perspectives. 5 GHz band provides faster data rates at a shorter distance, whereas 2.4 GHz band offers coverage for greater distances but supports lower rates. New technologies, particularly unlicensed LTE variants including LTE-U, LAA-LTE, and MulteFire, as mentioned in Chap. 1, have been targeted to operate in the 5 GHz band alongside Wi-Fi. The selection of the 5 GHz band for U-LTE technologies (rather than the 2.4 GHz band) is mainly due to the following reasons:

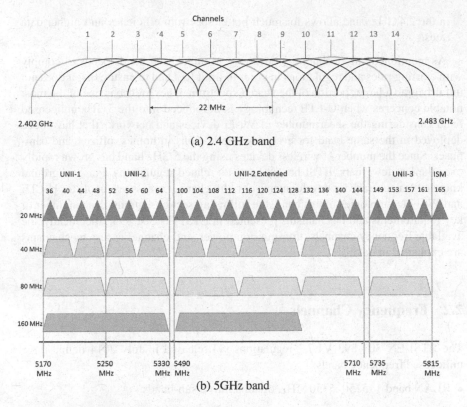

(a) 2.4 GHz band

(b) 5GHz band

Fig. 2.1 2.4 and 5 GHz unlicensed spectrums

- *More available channels*: In the 2.4 GHz band, only 14 channels, each of which provides 20 MHz of bandwidth, are defined. In USA (or Europe), only 11 (or 13) of those channels are legally available. However, those channels overlap excessively with one another. Due to this overlapping, the maximum possible number of parallel independent connections is limited to 3 channels (channels 1, 6, and 11). In the 5 GHz band, there are 21 nonoverlapping 20 MHz channels (or 9 nonoverlapping 40 MHz channels). Figure 2.1 depicts spectrum analyzer views of radio channels defined in 2.4 and 5 GHz bands, respectively.
- *Lower level of interference*: Since the 2.4 ISM band was released for Wi-Fi technology use more than fifteen years ago, this band is overcrowded with billions of existing Wi-Fi devices. There are also many consumer products which use this band, including microwave ovens, cordless phones, baby monitors, and garage door openers. In contrast, the relatively recent release of the 5 GHz band for private use makes this band much less crowded and thus having a much lower level of RFI.
- *Higher performance*: The 5 GHz band operates on a larger spectrum and does not suffer the overcrowding. Therefore, compared to the 2.4 GHz band, each channel

in the 2.4 GHz band allows for much better spectrum efficiency and higher data rates.

As mentioned, any technology operating in unlicensed bands needs to comply with existing rules and regulations in order to limit RFI and to ensure that it does not unfairly grab a larger portion of the shared spectrum. Coexistence is one of the most notable concerns when U-LTE technology is introduced into the 5 GHz unlicensed band considering the sheer number of Wi-Fi devices and networks that have been deployed in the same band for everyday applications in homes, offices, and campuses. Since the number of wireless devices using the 5 GHz band has grown rapidly over the last few years, ETSI has updated its related regulations. For background knowledge necessary for developments of radio channel access protocols for U-LTE and Wi-Fi technologies in this band, the following sections summarize a number of key requirements and mechanisms presented in ETSI EN 301 893. Specifically, the available frequency channels, transmission power, and channel access mechanisms are explained in detail.

2.2 Frequency Channels

The ETSI EN 301 893 V1.7.2 regulations [1] released in July 2014 define three unlicensed frequency bands:

- RLAN band 1: 5150–5350 MHz, divided into 2 sub-bands

  - Sub-band I: 5150–5250 MHz. This sub-band is comparable to FCC U-NII-1.
  - Sub-band II: 5250–5350 MHz. This sub-band is comparable to FCC U-NII-2.
  - RLAN band 2: 5470–5725 MHz. This band comparable to FCC U-NII-2 extended (U-NII-2e).
  - RLAN band 3, also known as Broadband Radio Access Networks (BRAN): 5725–5875 MHz. This sub-band is comparable to FCC U-NII-3 (5725–5825 MHz) band with a higher upper frequency range.

The radio channels defined in the 5 GHz band by ETSI 301 893 standard (with a reference to FCC regulations) are summarized in Fig. 2.1b. Technical details and the availability of each channel in four main regions (USA, Europe, Japan, and China) are presented in Fig. 2.2.

2.3 Transmission Power

Each of the bands defined by ETSI EN 301 893 V1.7.2 regulations [1] has different maximum allowable transmission power levels. Note that the RF output power is defined as the mean Equivalent Isotropically Radiated Power (EIRP) of the equipment during a transmission burst. In general, the limits are valid for the device with antenna gain and cable loss and not only the output power of WLAN module.

Channel Number	Center Frequency	U.S	Europe	Japan
36	5180	Yes	Indoors	Yes
38	5190	No	No	Client Only
40	5200	Yes	Indoors	Yes
42	5210	No	No	Client Only
44	5220	Yes	Indoors	Yes
46	5230	No	No	Client Only
48	5240	Yes	Indoors	Yes
52	5260	DFS	Indoors/DFS/TPC	DFS/TPC
56	5280	DFS	Indoors/DFS/TPC	DFS/TPC
60	5300	DFS	Indoors/DFS/TPC	DFS/TPC
64	5320	DFS	DFS/TPC	DFS/TPC
100	5500	DFS	DFS/TPC	DFS/TPC
104	5520	DFS	DFS/TPC	DFS/TPC
108	5540	DFS	DFS/TPC	DFS/TPC
112	5560	DFS	DFS/TPC	DFS/TPC
116	5580	DFS	DFS/TPC	DFS/TPC
120	5600	No	DFS/TPC	DFS/TPC
124	5620	No	DFS/TPC	DFS/TPC
128	5640	No	DFS/TPC	DFS/TPC
132	5660	DFS	DFS/TPC	DFS/TPC
136	5680	DFS	DFS/TPC	DFS/TPC
140	5700	DFS	DFS/TPC	DFS/TPC
149	5745	Yes	SRD	No
153	5765	Yes	SRD	No
157	5785	Yes	SRD	No
161	5805	Yes	SRD	No
165	5825	Yes	SRD	No

Fig. 2.2 Details of 5 GHz unlicensed channels in different regions

2.3.1 RLAN Band 1 (5150–5350 MHz)

2.3.1.1 Indoor-Only Sub-band I (5150–5250 MHz)

The first RLAN sub-band includes the channels 36–48 and has an EIRP power limit to 23 dBm (200 mW). These channels are considered for indoor-only usage and do not require any Dynamic Frequency Selection (DFS) or Transmit Power Control (TPC) features.

2.3.1.2 Indoor-only Sub-band II (5250–5350 MHz)

In the second sub-band of the RLAN band 1 with channels 52 to 64, ETSI has set the EIRP power limit to 23 dBm (200 mW) for devices with TPC and 20 dBm (100 mW) for devices without TPC. For a device with TPC, the mean EIRP at the lowest power

level of the TPC range must not exceed 17 dBm (50 mW). This band requires DFS
support (and requires TPC support in Europe and Japan).

2.3.2 RLAN Band 2 (5470–5725 MHz)

Channels from 100 to 140 are part of the second RLAN band and have an EIRP
power limit of 30 dBm (1000 mW) for devices with DFS and TPC support, 27 dBm
(500 mW) for non-TPC devices, and 20 dBm (100 mW) for devices with neither TPC
or DFS support. The mean EIRP power level for a slave device with TPC must not
exceed 24 dBm at the lowest TPC power level if the device is also capable of radar
detection or 17 dBm otherwise.

2.3.3 BRAN (5725–5875 MHz)

ETSI has restricted the channels 155–171 for Broadband Wireless Access (BWA) use
only. The idea is to provide Internet access to locations without any wired access network available. The maximum EIRP output power has been set to 36 dBm (4000 mW)
with the limitation of RF power into antenna of 30 dBm (1000 mW).

2.4 Transmission Power Control

Dynamic adjustment of the transmission power is intended to reduce RFI. Dynamically adjusting the transmission power facilitates the shared use of the
5250–5350 MHz and 5470–5725 MHz frequency bands with satellite services. TPC
determines the minimum transmission power necessary to maintain the connection
with the partner (such as an access point).

If TPC is not used within these frequency bands, the highest permissible average
EIRP and the corresponding maximum EIRP density are reduced by 3 dB. This
restriction does not apply to the frequency range of 5150–5350 MHz. Without DFS
and TPC, a maximum of only 30 mW EIRP is permitted. When DFS and TPC are
used, a maximum 1000 mW EIRP is permitted as the transmission power (compared
with 100 mW with 802.11 b/g, 2.4 GHz, DFS and TPC are not possible here). The
higher maximum transmission power not only compensates for the higher attenuation
of 5 GHz radio waves in air, but also makes noticeably longer ranges possible than
in the 2.4 GHz range.

2.5 Dynamic Frequency Selection

DFS was stipulated to (i) detect interference from radar systems (radar detection) and to avoid co-channel operation with these systems and (ii) to provide, on aggregate, a near-uniform loading of the spectrum (Uniform Spreading). DFS is stipulated for the frequency ranges of 5250–5350 MHz and 5470–5725 MHz. It is optional for the frequency range of 5150–5250 MHz.

DFS initially assumes that no channel is available in the corresponding frequency band. The WLAN device selects an arbitrary channel at the start and performs what is known as a Channel Availability Check (CAC). Before transmitting on a channel, a Channel Observation Time (COT) of 60 s is observed to allow a check to see whether a different device is already working on this channel and the channel is therefore occupied. If the channel is occupied, then a different channel is checked by the CAC. If not, then the WLAN device can perform its transmission operation. Even during operation, a check is run to see whether a primary application such as a radar device is using this channel. This exploits the fact that radars frequently work according to the rotation method, whereby a tightly bundled directional transmission signal is transmitted by a rotating antenna. A remote receiver perceives the radar signal as a short pulse (radar peak). If a device receives such a radar peak, then it pauses the transmission operation and monitors the channel for further pulses. If additional radar peaks occur during the COT, then a new channel is selected automatically. A check of this type is required to be carried out every 24 h. This is why interrupting the data transmission for 60 s is unavoidable.

2.6 Channel Access Mechanisms

In order to avoid channel collisions when two or more than two devices transmit the signal in the same channel at the same time, Listen-Before-Talk (BLT) strategy is employed. ETSI EN 301 893 V1.7.2 [1] describes two mechanisms that require equipment or a device to apply CCA before using the channel. The first mechanism is frame based equipment (FBE) which defines a fixed (not directly demand-driven) timing frame for channel access. The second mechanism is load based equipment (LBE) which defines demand-driven timing frame.

2.6.1 FBE-Based Mechanism

A simplified flowchart and an illustrative example of the channel access procedure used for FBE are given in Figs. 2.3 and 2.4, respectively.

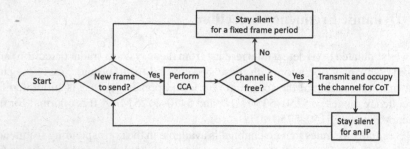

Fig. 2.3 Simplified flowchart of FBE

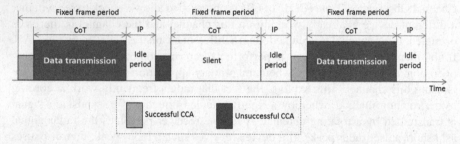

Fig. 2.4 An illustrative example of FBE

FBE shall comply with the following requirements:

- *R1*: Before starting transmissions on an operating channel, the equipment shall perform a CCA check using energy detect (ED). The equipment shall observe the channel for the duration of the CCA observation time. The operating channel shall be considered occupied if the energy level in the channel exceeds the threshold corresponding to the power level.
- *R2*: If the CCA procedure finds that the channel is clear, the equipment may transmit immediately and occupy the channel for a fixed time period.
- *R3*: If the CCA procedure finds that the channel is occupied, the equipment shall not transmit on that channel during the next fixed frame period.
- *R4*: The total time during which the equipment has transmissions on a given channel without re-evaluating the availability of that channel is defined as the *Channel Occupancy Time* (CoT).
- *R5*: After occupying the channel for CoT, the equipment keeps silent and waits for a short time, namely *Idle Period* (IP).
- *R6*: Toward the end of the idle period, the equipment shall perform a new CCA procedure as described in R1 above.
- *R7*: The equipment, upon correct reception of a packet which was intended for this equipment, can skip CCA and immediately proceed with the transmission of management and control frames, e.g., acknowledgment (ACK) and block ACK frames.

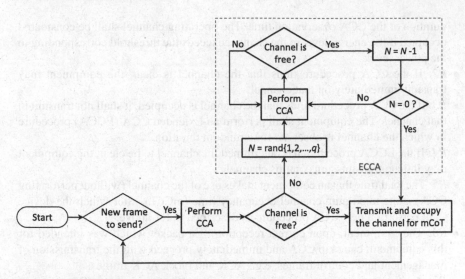

Fig. 2.5 Simplified flowchart of LBE

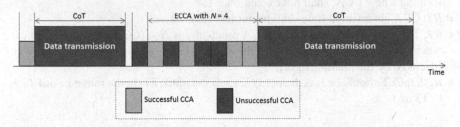

Fig. 2.6 An illustrative example of LBE

- *R8*: A consecutive sequence of such transmissions by the equipment, without it performing a new CCA, shall not exceed the maximum CoT.
- *R9*: CCA observation time shall be not less than 20 μs.
- *R10*: CoT shall be in the range from 1 to 10 ms.
- *R11*: The minimum IP shall be at least 5% of CoT used by the equipment for the current fixed frame period.

2.6.2 LBE-Based Mechanism

A simplified flowchart and an illustrative example of the channel access procedure used for LBE are shown in Figs. 2.5 and 2.6, respectively.
LBE shall comply with the following requirements:

- *R1*: Before starting transmissions on an operating channel, the equipment shall perform a CCA check using ED. The equipment shall observe the channel for the duration of the CCA observation time. The operating channel shall be considered occupied if the energy level in the channel exceeds the threshold corresponding to the power level.

- *R2*: If the CCA procedure finds that the channel is clear, the equipment may transmit immediately on that channel.
- *R3*: If the CCA procedure finds that the channel is occupied, it shall not transmit in that channel. The equipment shall perform an Extended CCA (ECCA) procedure in which the channel is observed for a random duration.
- *R4:* If the ECCA procedure has determined the channel to be clear, the equipment may initiate transmissions on this channel.
- *R5*: The total time that an equipment makes use of the channel (without performing CCA) is the maximum channel occupancy time (mCoT), after which the device shall perform a new CCA procedure as described in R1 above.
- *R6*: The equipment, upon correct reception of a packet which was intended for this equipment, can skip CCA and immediately proceed with the transmission of management and control frames, e.g., ACK and block ACK frames.
- *R7*: A consecutive sequence of transmissions by the equipment, without it performing a new CCA, shall not exceed mCoT.
- *R8*: CCA observation time shall be not less than $20\,\mu$s.
- *R9*: The random duration in an ECCA procedure is $N \times$ (CCA observation time), where N is randomly selected in the range $\{1, 2, \ldots, q\}$, $q \in \{4, 5, \ldots, 32\}$ (declared by the manufacturer).
- *R10*: mCoT should be less than $(13/32) \times q$ ms (mCoT is in the range from 1.625 to 13 ms).

Reference

1. *ETSI EN 301 893 V1.7.2 (2014-07): Broadband radio access networks (BRAN); 5 GHz high performance RLAN; Harmonized EN covering the essential requirements of article 3.2 of the R&TTE Directive*, European Telecommunications Standards Institute Std., 2014.

Chapter 3
LTE-Advanced: An Overview

This chapter provides a high-level overview of LTE-Advanced (LTE-A) networks and associated technologies to form a basis for discussion of the coexistence issues that exist for unlicensed LTE and Wi-Fi. Understanding the underlying architecture and protocols employed in LTE-A networks will provide a comparative framework to grasp how, and at what levels, LTE and Wi-Fi networks may interact and interfere with each other, and form a greater understanding of the challenges to be address in designing coexistence mechanisms. Specifically, this chapter will overview the LTE-A network, as well as its capabilities and protocols, with specific emphasis on the physical layer and medium access sub-layers to illuminate specific sources of coexistence issues. Proposed changes which may be included in future LTE releases are discussed in the context of LTE/Wi-Fi coexistence.

3.1 System Overview

The enhancements to the Long Term Evolution/System Architecture Evolution (LTE/SAE) to meet the requirements set out for fourth generation (4G) cellular networks are collectively known as LTE-Advanced (LTE-A). The LTE-A requirements were formalized by the 3rd Generation Partnership Project (3GPP) in LTE releases 10 through 13 [8]. LTE itself was a logical evolution from the technologies used in previous generations in order to meet the increasing demands for higher data rates and improved quality of service. LTE met these demands at the access level through increased spectral efficiency and improved mobility support and cell edge data rates. The increased spectral efficiency was achieved by using orthogonal frequency division multiple access (OFDMA) and single-carrier frequency division multiple access (SC-FDMA) in the downlink (DL) and uplink (UL), respectively. Improvements in mobility support and cell edge data rates were achieved through enhanced adaptive modulation and bandwidth selection and DL spatial multiplexing and multiple input/multiple output support. Beyond the access layer, LTE transitioned to an all-IP packet switched core network with the introduction of the evolved

© The Author(s) 2017
Q.-D. Ho et al., *Long Term Evolution in Unlicensed Bands*,
SpringerBriefs in Electrical and Computer Engineering,
DOI 10.1007/978-3-319-47346-8_3

packet core, and a flattened network architecture of enhanced base stations called evolved NodeB's (eNB) interconnected via high-speed data links. Combined, these fundamental changes to the cellular network architecture have allowed LTE networks to significantly increase user data rates and reduce control and user plane latency and connection set-up and handover times. LTE-A represents the further, and ongoing, evolution of cellular networks to continue to meet the ever-increasing demands for higher data rates, user mobility support, and efficient support of a growing number of wireless devices.

3.1.1   Network Architecture

The requirements to provide high data rates while supporting high-speed mobility requires the ability to set up and tear down user connections and manage inter-cell handoffs with as little latency as possible. In previous generations of cellular networks, a hierarchical structure consisting of base stations or NodeBs connected to a central controller had been used. This star or cluster architecture requires additional hops in both data transmissions and hand off negotiation which can introduce significant delay. Controllers were responsible for managing all data and control traffic as well as handoffs betweens several pairs of base stations. For many increasingly ubiquitous end-user applications, such as online gaming and voice/video over the internet, the additional latency in connection set up and handover can impair the user-perceived quality of experience.

The flat architecture adopted by LTE networks is depicted in Fig. 3.1. The migration of local functions to eNBs and global functions to the EPC were driven by the requirements of reduced latency and higher data rates. The functions of radio network and medium access control, handoff requests, negotiations, and management, as well as some other truly local functions, are migrated to the eNBs [2]. The eNBs are interconnected via the low-latency X2 interface in a mesh configuration to allow for fast user handover, including forwarding of queued data for seamless user experience. Additionally, with direct connections between neighboring cells, this architecture facilitates more effective multi-point transmission, coordination, and inter-cell interference and load management, independent of conditions in other areas of the network. The global functions and connections to external networks are handled at the evolved packet core (EPC). The functional split between the various components of the network, as well as the implementation of the necessary layers of the network protocol stack, is shown in further detail in Fig. 3.2. In addition to those functions listed in the figure, the mobile management entity (MME) handles authentication, authorization and accounting functions. The packet data network gateway (P-GW) and serving gateway (S-GW) handle user data packet forwarding, filtering, and usage tracking, as well as acting as a mobility anchor for inter-eNB and inter-RAT handovers. Further, the distributed radio network and resource management and medium access control allows eNBs to quickly adapt to changing radio medium conditions and provide timely user scheduling based on local information.

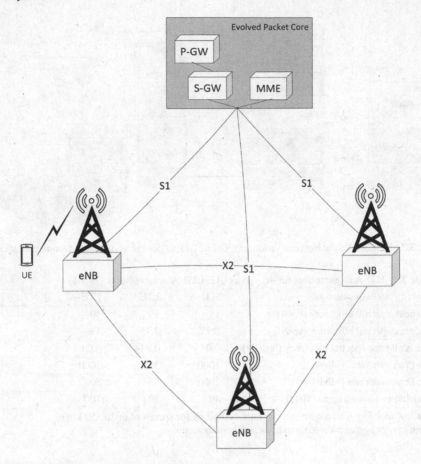

Fig. 3.1 Basic structure of LTE cellular networks

3.1.2 Capabilities and Features

While the gains made by LTE were significant, they fell short of the requirements set out for 4G networks by the International Telecommunications Union, specifically in the case of peak data rates, spectral efficiency, and cell edge performance [10]. The continuing evolution which became LTE-A was finally able to achieve the necessary targets to meet the ITU requirements for 4G. Some important ITU requirements, and achieved performance levels for LTE and LTE-A, are highlighted in Table 3.1.

Among other innovations, LTE-A extended bandwidth scalability in LTE by supporting carrier aggregation, both within and across frequency bands. Discontiguous aggregation is supported to ensure a higher bandwidth is available for providers who cannot support it in contiguous spectrum allotments, allowing the development of license-assisted access (LAA-LTE) into the unlicensed and TV whitespace bands. Backwards compatibility is maintained by using bandwidths for each carrier

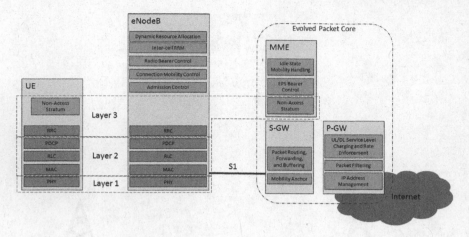

Fig. 3.2 Functional split between various entities in LTE under the system architecture evolution

Table 3.1 ITU-A requirements for 4G versus LTE/LTE-A achievements [7–10]

Description/Requirements	ITU-A	LTE	LTE-A
DL peak spectral efficiency (bps/Hz)	15	15	30
UL peak spectral efficiency (bps/Hz)	6.75	3.75	15
Min. cell edge spectral efficiency (bps/Hz)	0.04	0.024	0.04
DL Peak data rates (Mbps)	1000[a]	300	1000
UL Peak data rates (Mbps)	1000[b]	75	500
Scalable bandwidth up to (MHz)	40	20	100[b]

[a]For low mobility with requirement of min. 100 Mbps for speeds of up to 350 km/h.
[b]With carrier aggregation of up to five carrier components

component which match those used in LTE. LTE-A also expands MIMO/spatial multiplexing support up to 8×8 for DL and 4×4 for UL, adds coordinated multi-point operation to increase spectral efficiency and cell edge data rates, and improves heterogeneous network planning with the enhancement of support for small cells and relay nodes to increase area coverage with reduced power requirements.

3.2 Channel Access Mechanisms

Like other cellular access technologies, LTE-A has been designed for use on dedicated licensed spectrum allocations where there is, generally, no need to contend for channel access. While interference, fading, and path loss can corrupt LTE transmissions, and recovery and retransmission functions are necessary, in general a centrally controlled and tightly scheduled channel access mechanism is able to guarantee service levels required by all UL and DL traffic [2]. UL/DL separation

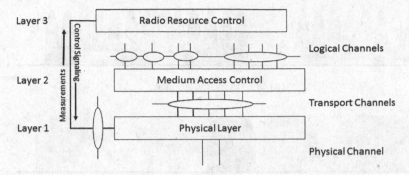

Fig. 3.3 E-UTRA radio interface protocol architecture

is achieved through either time-division duplexing (TDD) or frequency-division duplexing (FDD). Orthogonal frequency-division multiple access (OFDMA) is used in the DL, allowing the eNB to efficiently schedule transmissions for many users in the same transmission time interval. Single-carrier frequency-division multiple access (SC-FDMA) is used in the uplink in order to reduce the power consumption requirements of battery-dependent user equipment (UE) to communicate with the eNB. Further, coordinated multipoint (CoMP) is supported by allowing UEs to be configured to process channel state information (CSI) from multiple eNBs, and both single-user (SU) and multi-user (MU) MIMO are supported in multiple configurations to achieve transmit diversity or multi-layer transmissions with beamforming possible in both horizontal and vertical dimensions.

The basic structure of the protocol stack used in LTE networks to facilitate channel access is shown in Fig. 3.3 [6]. UL and DL transmissions are divided amongst several physical channels, according to the type of transmission, i.e. user traffic or control information, and the type of transmission, i.e. broadcast or unicast and scheduled or random access (random access is primarily used by a UE which has not yet associated to an eNB). The information-bearing physical channels are mapped by the PHY layer into transport channels supplied to the MAC sublayer, which in turn remaps these into several logical channels provided to the higher layers.

3.2.1   LTE-A Physical Layer

The LTE PHY layer is designed to be both highly adaptable as well spectrally and power efficient. In the DL, OFDMA is used to schedule many signals in the same transmission time interval and achieve a high spectral efficiency; however, the very high peak to average power ratio makes this multiple access strategy unattractive in the UL for battery-dependent devices [5, 6]. SC-FDMA is used in the UL to maintain a satisfactory spectral efficiency while significantly improving power efficiency and battery life. Example carrier allocations for both DL and UL are shown in Fig. 3.4.

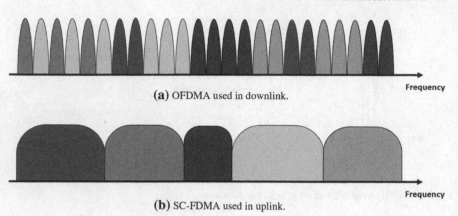

(a) OFDMA used in downlink.

Frequency

(b) SC-FDMA used in uplink.

Frequency

Fig. 3.4 Multiple access strategies used in LTE for uplink and downlink

Each color and shade represents the carriers allocated to a specific user. The assignment of sub-carriers to a given UE are driven by the specific loss and interference experienced by that user, called channel state information (CSI), in order to maximize the overall achieved rate of all users. As shown in Fig. 3.4a, in the DL LTE-A supports both localized (contiguous) and distributed allocations, to best use CSI to maximize network efficiency. Sub-carrier assignments can further span disjoint frequency bands, including into unlicensed bands. Figure 3.4b depicts an example UL allocation where a given sub-carrier, with variable bandwidth, is assigned to a single UE, again based on CSI. It should be emphasized that Fig. 3.4 shows a single instant of time, over a subset of the available carriers in the band.

All transmissions are organized into frames comprised of twenty slots, every two of which comprise a subframe [5]. An example of one configuration for the LTE frame structure is shown in Fig. 3.5. A single sub-carrier, with either 7.5 or 15 kHz bandwidth, paired with a transmission duration required to transmit a single OFDM symbol is called a resource element. The minimum unit which can be allocated to a physical channel or user is the physical resource block (PRB). For simplicity, in the breakout, a single PRB is depicted, though a single PRB spans only a small subset of the available sub-carriers. Depending on the configuration used, a PRB is formed of either 3, 6, or 7 OFDM symbols and 12 or 24 sub-carriers allocated for one 0.5 ms slot. Between 6 and 10 PRBs will then be allocated to a UE in order to achieve bandwidths between 1.4 and 20 MHz (though the occupied bandwidth will be smaller, 1.08–18 MHz). Higher bandwidths are achieved through carrier aggregation in the same slot, either contiguous or not.

Two distinct frame structures are defined to support both time division duplexing (TDD) and frequency division duplexing (FDD), as well as a third frame structure specifically for license assisted access (LAA-LTE). All three frame types have the same basic structure shown in Fig. 3.5, with the differences being in how transmission are scheduled. Both half- and full-duplex are supported in FDD and all 10 subframes are available for both UL and DL. In half-duplex configurations, transmission are

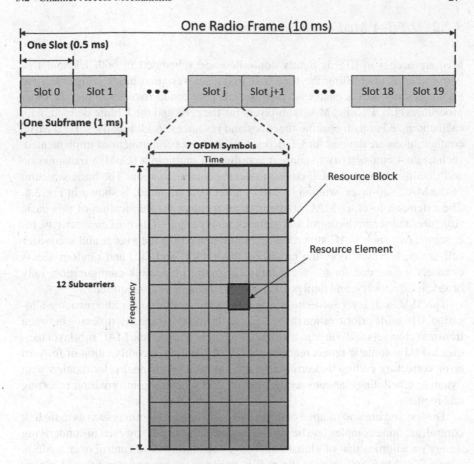

Fig. 3.5 LTE frame and resource block structure

separated in both time and frequency, while for full-duplex transmissions separation is in frequency only. For TDD, the frame is organized into two 5 ms half-frames, with several UL/DL configurations and switching patterns supported. As of Release 13, the special frame structure defined for LAA-LTE reserves all 10 subframes for DL, with transmission able to occupy one or more consecutive subframes, but required to start somewhere within the first subframe. LAA-LTE is expected to be expanded to support both UL and DL in future releases.

Beyond the features discussed in detail, the LTE-A PHY layer provides forward error correction and automatic repeat request functions, modulation/demodulation of physical signals, mapping and rate matching of physical channels to transport channels, frequency and time synchronization, and MIMO antenna processing, including transmit diversity and beamforming. A variety of modulation schemes are supported depending on channel conditions, distance from receiver, and power requirements (up to 256 QAM in the DL).

3.2.2   LTE-A Medium Access Control

Medium access in LTE is tightly controlled and scheduled in both UL and DL, with the eNB controlling the time and frequency resource block assignment for all but random access channels (used for UE connection requests and some other procedures) [3]. Distinct MAC sub-layers for the eNB and the UE are defined, with each optimized to their specific functions and resources. Additionally, several MAC configurations are defined for UE depending on the specific functions implemented, such as dual connectivity to support coordinated multipoint (CoMP) transmission and sidelink channels for UE device-to-device communication. The basic structure of the MAC sub-layer, without CoMP or sidelink configured, is shown in Fig. 3.6. The extension to other MAC configurations requires the duplication of this basic structure, and separate control and traffic channels required. In dual connectivity, for example, two such MAC structures are implemented for the master and secondary cell group; however, only the broadcast, shared UL and DL, and random access channels are needed for the secondary cell group. In sidelink configuration, only broadcast, discovery, and duplex UL/DL shared channels are required.

The MAC sub-layer facilitates reliable data transfer through radio resource allocation, UL traffic prioritization in the UE, and the mapping and multiplexing between transport channels and one or more logical channels. Further, the MAC sub-layer handles hybrid automatic repeat requests (HARQ) signaling, a combination of forward error correcting coding and error detection, as well as channel prioritization with dynamic scheduling, random access control, and scheduling information reporting and requests.

The specific implementation of these, and other layer 2 functions such as radio link control, are quite complex and beyond the scope of this brief; however, the underlying design paradigm is that of global knowledge facilitating tight control over medium access. The MAC sub-layer in the eNB is aware of the channel conditions for each UE to be scheduled in both the UL and DL, and the UE adheres to the schedule of time

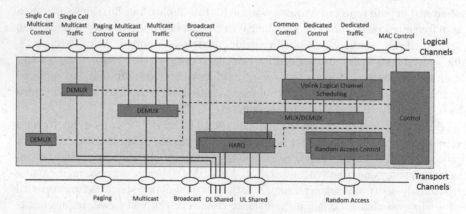

Fig. 3.6 LTE MAC sublayer structure and channel mapping

and frequency provided by the eNB. PRBs are assigned to UL and DL transmissions to achieve the greatest advantage from local channel conditions. Little consideration of inter-user or inter-carrier interference is necessary, due to the dedicated frequency bands and orthogonal carriers discussed in Sect. 3.2.1. As a result, LTE-A is able to achieve high spectral efficiency and reliability.

3.3 Changes Expected for Future Releases

LTE-A brings cellular networks into the realm of 4G, as defined by [10]. As far as LTE-A has taken us, it will not be enough for fifth generation (5G) networks which are expected to support existing and new use cases ranging from smart cities and Internet of Things devices (massive machine type communication) to self-driving vehicles and industrial automation (ultra-reliable low latency communications) with high speed mobile broadband on the order of giga*bytes* per second [11]. Some of the specific requirements are outlined in Table 3.2. Beyond these specific targets set by the ITU, 5G networks must also achieve $10 \times$ reduced latency, $3 \times$ improved spectral efficiency and $100 \times$ network energy efficiency, compared to 4G networks.

In order to move towards 5G networks, 3GPP has numerous study items underway and planned for future releases, to meet the ITU requirements in [11]. These include significant enhancements to inter- and intra-band carrier aggregation and licenses-assisted access to ISM bands, TV white space, and other under-utilized spectrum resources, as well as multi-carrier enhancements and improved CoMP and device-to-device communications [1]. Wireless network virtualization and hardware resource sharing, cloud-based radio access networks, and new system architectures are under consideration to support the requirements around energy efficiency and low-latency communications, among others. Additionally, the possibility of moving to entirely new radio and medium access protocols is also under consideration, which would not be hindered by the need to be backwards compatible, and move forward with only the necessity of meeting the ITU 5G requirements.

Table 3.2 ITU-A requirements for 4G versus ITU-2020 requirements for 5G [11]

Description/Requirements	ITU-A	ITU-2020
Peak data rates (Gbps)	1	20
Average user data rates (Mbps)	1	100
Mobility support(km/h)	350	500
Connection density (devices/km^2)	10^5	10^6
Traffic capacity (Mbits/s/m^2)	0.1	10

References

1. 3GPP, "3GPP work programme (and associated reference links therein)," July 2016. Accessed: July 12, 2016 [Online]. Available: www.3gpp.org/dynareport/GanttChart-Level-2.htm
2. 3GPP, "Evolved universal terrestrial radio access (E-UTRA) and evolved universal terrestrial radio access network (E-UTRAN); Overall description," *3GPP TS 36.300 version 13.3.0 Release 13*, 2016.
3. 3GPP, "Evolved universal terrestrial radio access (E-UTRA); Medium access control (MAC) protocol specification," *3GPP TS 36.321 version 13.1.0 Release 13*, 2016.
4. 3GPP, "LTE-Advanced," Whitepaper, June 2013. Accessed: June 21, 2016 [Online]. Available: www.3gpp.org/technologies/keywords-acronyms/97-lte-advanced
5. 3GPP, "LTE; Evolved universal terrestrial radio sccess (E-UTRA); Physical channels and modulation," *3GPP TS 36.211 version 13.1.0 Release 13*, 2016.
6. 3GPP, "LTE; Evolved universal terrestrial radio access (E-UTRA); Physical layer; General description," *3GPP TS 36.201 version 13.1.0 Release 13*, 2016.
7. 3GPP, "LTE," Whitepaper, ND. Accessed: June 21, 2016 [Online]. Available: www.3gpp.org/technologies/keywords-acronyms/98-lte
8. 3GPP, "Requirements for further advancements for Evolved Universal Terrestrial Radio Access (E-UTRA) (LTE-Advanced)," *3GPP TR 36.913 version 13.0.0 Release 10*, 2011.
9. M.F.L. Abdullah and A.Z. Yonis, "Performance of LTE release 8 and release 10 in wireless communications," in *Cyber Security, Cyber Warfare and Digital Forensic (CyberSec), 2012 International Conference on*, June 2012, pp. 236–241.
10. ITU, "IMT Vision Framework and overall objectives of the future development of IMT for 2020 and beyond," *Rep. ITU-R M.2083-0*, 2015.
11. ITU, "Requirements related to technical performance for IMT-Advanced radio interface(s)," *Rep. ITU-R M.2134*, 2008.

Chapter 4
IEEE 802.11/Wi-Fi Medium Access Control: An Overview

In order to understand and analyze the interactions between U-LTE and Wi-Fi when sharing the 5 GHz frequency band, knowledge of the architectures and medium access control (MAC) mechanisms currently adopted by these two technologies is required. This chapter focuses on the Wi-Fi technology and begins with an overview of the evolution of IEEE 802.11/Wi-Fi. Five existing generations, as well as the next generation, of IEEE 802.11/Wi-Fi are presented. The majority of this chapter provides the underlying ideas and detailed mechanisms of the CSMA/CA MAC protocol. Important observations on how CSMA/CA senses and occupies the radio medium when the LTE network is operating in the vicinity are highlighted.

4.1 IEEE 802.11/Wi-Fi Evolution

The IEEE 802.11 standard is a branch of 802 family of standards created and maintained by the Institute of Electrical and Electronics Engineers (IEEE) Local Area Network (LAN)/Metropolitan Area Network (MAN) Standards Committee (IEEE 802). It defines a set of specifications of physical (PHY) and MAC layers of Wireless Local Area Networks (WLANs) operating in a number of unlicensed radio frequency bands including sub-Ghz, 2.4, 5, and 60 GHz. Commercial products using IEEE 802.11 standards are branded as Wi-Fi. The Wi-Fi Alliance is an organization made up of leading wireless equipment and software providers with a mission of certifying all 802.11-based products for interoperability and promoting the term Wi-Fi as the global brand name [6]. According to the statistics made by Wi-Fi Alliance, in January 2016 about 12 billion Wi-Fi units have been shipped and deployed in homes, offices, buildings, factories, and so on. As a result, Wi-Fi has become one of the most prolific technologies around the world.

The first IEEE 802.11 standard was introduced in 1997 [1]. Since then, this technology has evolved with different generations to meet the increasing demands in system throughput and to support a wider range of features and applications. The most commonly deployed standards are 802.11a, 802.11b, 802.11g, 802.11n,

© The Author(s) 2017
Q.-D. Ho et al., *Long Term Evolution in Unlicensed Bands*,
SpringerBriefs in Electrical and Computer Engineering,
DOI 10.1007/978-3-319-47346-8_4

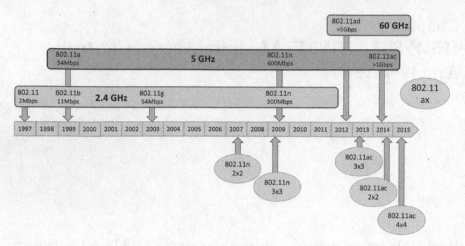

Fig. 4.1   IEEE 802.11/Wi-Fi standard evolution

and 802.11ac. Today, most businesses are using 802.11n and are looking to adopt 802.11ac as it is the fastest and latest available. Figure 4.1 sketches the evolution of IEEE 802.11/Wi-Fi technology over the past 20 years [1, 4, 5]. The original 802.11 standard (introduced in 1997) can support only 1 or 2 Mbps, which is quite low by modern standards. It was aimed to provide an alternative to and/or replace wired ethernet connectivity. The WLAN rate was then significantly improved in the IEEE 802.11b and IEEE 802.11a updates released in 1999. The 802.11b standard, considered as the first-generation WLAN technology, features two well-known spread spectrum technologies to distribute packets over a wireless medium, Frequency-Hopping Spread Spectrum (FHSS) and Direct-Sequence Spread Spectrum (DSSS), that are still in use by most wireless networks today. 802.11b operates in 2.4 GHz frequency band and can support a maximum data rate of 11 Mbps. The IEEE 802.11a standard— the second-generation WLAN technology—employs orthogonal frequency division multiplexing (OFDM) technology to enable a higher data rate of up to 54 Mbps. Using the 5 GHz band, this standard has a further significant advantage when compared to the relatively crowded 2.4 GHz band. The third-generation WLAN technology— IEEE 802.11g—was released in 2003. Also operating in the 2.4 GHz band (as does IEEE 802.11b), 802.11g uses OFDM to match the 54 Mbps data rate achieved by 802.11a. 802.11e was then introduced in 2005 as an enhancement to 802.11a and b with quality-of-service (QoS) support. 802.11e is capable of operating at radio frequencies of up to 5.8 GHz and is most suitable for networks with multimedia applications. In order to further boost the data rate, in 2007, the 802.11n using multiple-input–multiple-output (MIMO) technology was introduced. Considered as the fourth-generation WLAN, 802.11n uses both the 2.4 and 5 GHz frequency bands, can support up to 600 Mbps, and has now become the dominant WLAN standard.

The demand for higher throughput over the wireless medium escalated in the late 2000s, driven by mass adoption of smart phones, tablets, and video on demand services such as YouTube and Netflix. To address this, the 802.11ac standard, also known as the fifth-generation or gigabit Wi-Fi, was approved in 2014 [4]. It operates only in 5 GHz band, incorporating the enhanced air interface of 802.11n with wider bandwidth, more MIMO streams, and high-density modulation to support at least 1 Gbps. This standard is now incorporated into many mainstream Wi-Fi products. The 802.11ad is another gigabit Wi-Fi which was introduced in 2012, operating in the unlicensed 60 GHz band and offering much higher transfer rates than previous 802.11 standards (its theoretical maximum transfer rate is up to 7 Gbps). Technology based on the 802.11ad standard can supplement existing wireless networks for high-definition video streaming, offering the ability to off-load heavy demands on 2.4 and 5 GHz that other 802.11 standards are operating on. The 802.11ax—the successor to the 802.11ac and also called High-Efficiency Wireless (HEW)—is currently at an early stage of development and has the goal of providing 10 Gbps in both the 2.4 and the 5 GHz bands. This new standard implements several technologies to enhance the efficiency of channel utilization and is therefore expected to provide users with consistent and reliable data throughput in crowded wireless environments. One of the biggest enabling technologies of this efficiency is multi-user technology, in the form of both Multi-User MIMO (MU-MIMO) and Multi-User OFDMA (MU-OFDMA).

4.2 IEEE 802.11 CSMA/CA

Despite that fact that many generations of IEEE 802.11 have been developed to enhance data rates and support additional features as presented in the previous section, they basically share the same MAC mechanism which dictates how multiple devices can share the radio channel. In a nutshell, the 802.11 standard defines two operating modes: infrastructure and ad hoc [3]. In infrastructure mode, wireless clients are directly connected to an access point (AP) in a star topology. APs are connected to a distribution network, usually wired LANs, for Internet connection. In ad hoc mode, clients are connected to one another without any central AP. While ad hoc mode is only used in a limited number of scenarios, infrastructure mode is deployed in almost all Wi-Fi networks. The MAC layer for infrastructure Wi-Fi networks is composed of two radio channel coordination functions: distributed coordination function (DCF) and point coordination function (PCF).

4.2.1 DCF and PCF

DCF is a contention-based LBT mechanism called carrier-sense multiple access with Collision Avoidance (CSMA/CA) that works in an entirely distributed manner without any coordination [3]. With CSMA/CA, stations (STAs) independently perform carrier sensing and back-off procedures to compete for the channel access. DCF is a mandatory MAC function and implemented in all IEEE 802.11/Wi-Fi devices.

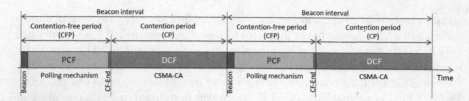

Fig. 4.2 PCF and DCF in IEEE 802.11

CSMA/CA is focused in this chapter and its technical details will be presented in
Sect. 4.2.3.

PCF is built on the top of DCF. It aims to support applications that require near-
real-time services. Basically, PCF splits the time into periodic intervals called beacon
intervals, each of which is composed of contention-free period (CFP) and contention
period (CP) [3]. CFP requires coordination from the access point (AP) and allocates
resources to STAs using polling mechanism. Specifically, AP maintains a list of
registered PCF-enabled STAs and polls each of them using CF-Poll frames. Only
after a STA is polled, it can start its data transmission. In case if the polled STA does
not have any frames to send, then it must transmit null frame. Channel access in CP
of PCF is handled by CSMA/CA protocol. PCF is specified as an optional MAC
function and has not been widely implemented due to its complexity. The timing
of PCF and DCF of IEEE 802.11 is sketched in Fig. 4.2. Within a given beacon
interval, the start and end of CFP are marked by beacon and CF-End control frames,
respectively. A CP follows each CFP and is terminated by a beacon frame signaling
the next beacon interval.

4.2.2 Basic Medium Access

The LBT mechanism employed by the IEEE 802.11/Wi-Fi CSMA/CA basically
follows the same philosophy of the carrier-sensing protocol family. When a STA
needs to transmit a new frame, the channel is sensed and if it is found idle the
frame is transmitted immediately. This simple mechanism is very effective when
the medium is not heavily loaded since it allows STAs to transmit with a minimum
delay. However, it cannot prevent channel access collisions when multiple STAs
detect a free channel and decide to transmit their frames at the same time. As a
result, in addition to this basic channel access, a number of important mechanisms
are mandated in CSMA/CA.

4.2.3 Medium Access with Collision Avoidance

Since it is difficult to detect collisions at a wireless receiver, the IEEE 802.11 protocol
tries to avoid collisions rather than detect and recover from collisions. This means

that CA mechanisms are mandated to reduce the collision probability at the points where collisions would most likely occur. Specifically, most collisions happen when the medium has become idle (as indicated by CS function) after a busy state. Several STAs could have been waiting for the medium to be available again, and then all transmit at the same moment the medium is detected free. This situation necessitates a "random" back-off procedure to resolve medium contention conflicts. Also, the use of inter-frame spaces (IFSs) of various durations helps to resolve the problem.

A flowchart detailing the CSMA/CA protocol is shown in Fig. 4.3 [3]. When a STA needs to transmit a new frame, if the channel has been continuously free over a Distributed IFS (DIFS) interval, it transmits immediately. Otherwise, STA defers its transmission until the channel becomes available. Then if the channel is detected to be continuously free over a Distributed IFS (DIFS) interval, the STA will initiate the back-off procedure to further defer its transmission over a random time interval. The back-off procedure starts with the selection of a random "slotted" back-off interval $BI_{slots} = \text{rand}[0, W]$, where $\text{rand}[0, W]$ is a random number uniformly distributed in the range from 0 to W, and W is the back-off window (when the system is started W is assigned to its minimum value W_{min}). Next, back-off counter w is initialized with BI_{slots} and decreased every time the medium is idle over a slot time (ST). This counter is frozen when a transmission is detected on the medium, and resumed when the channel is detected idle again for a DIFS interval. As soon as w finally reaches zero, the STA transmits its frame. It is important to note that this back-off procedure randomizes the channel access among STAs and thus helps to reduce the chance of collision. It also gives all STAs their fair share of the channel.

The destination STA, upon receiving a frame correctly, waits for a Short IFS (SIFS) interval immediately after the reception has completed and transmits an ACK

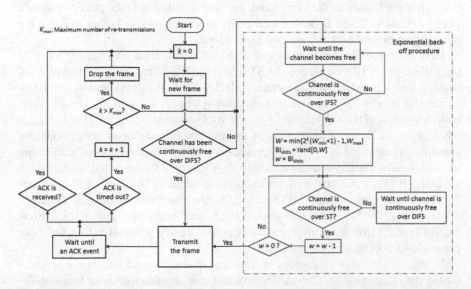

Fig. 4.3 Simplified flowchart of CSMA/CA

frame back to the source STA in order to confirm the correct reception. SIFS is the smallest IFS to give the highest priority channel access to ACK frames. If the source STA receives a confirmation, transmissions of the second and subsequent frames of a fragment burst will use SIFS instead of DIFS. Otherwise, the source STA activates the retransmission procedure for the lost frame.

When a transmission is lost (due to channel collision when two or more STAs decrease their back-off counter to zero at the same time and transmit their frames at the same time or transmission errors), the contention window W is doubled and applied for the retransmissions until it reaches a maximum value W_{max}. For the retransmissions, the back-off procedure is activated after the channel remains idle for an Extended IFS (EIFS) interval. When a frame transmission is successful, contention window W is reset to its minimum value W_{min}. When a maximum number of frame retransmissions is exhausted, the frame is discarded and W is also reset to its minimum value W_{min}.

The reason behind the exponential growth of contention window W is explained as follows: When a STA experiences a collision, it has no information on how many STAs are involved in the collision. If there are only few colliding frames, it would make sense to choose the random back-off interval from a small set of small values; i.e., W is small. But if many STAs are involved in a collision, then it makes sense to choose the back-off interval from a larger, more dispersed set of values; i.e., W is large. Otherwise, if several STAs select the back-off interval from a small set of values, more than one STA would choose the same back-off value with high probability. This will result in high probability of collision.

Figures 4.4 and 4.5 demonstrate the operations of the back-off procedure in two typical scenarios. As shown in Fig. 4.4, by randomly selecting back-off intervals, STAs C and D randomize their channel access to minimize the chance that they transmit their frames at the same time. In case a collision takes place, as shown in Fig. 4.5, STAs C and D double their contention windows to further increase the randomness in their back-off interval generations.

Here are some illustrative values of CSMA/CA operation parameters: ST = 20 μs, SIFS = 10 μs, DIFS = SIFS + 2 × ST = 50 μs, EIFS = Transmission time of ACK frame at lowest physical mandatory rate + SIFS + DIFS, $W_{min} = 31$, and $W_{max} = 1023$. Contention window of the initial transmission attempt is $W(0) = W_{min} = 31$. Contention window of the k-th retransmission is $W(k) = \min\{2^k(W_{min} + 1) - 1, W_{max}\}$, where $k \in \{1, 2, \ldots, K_{max}\}$, K_{max} is the maximum number of re-transmission attempts. Assuming $K_{max} = 7$, the progression of contention window with frame transmission/retransmissions is as follows: $W(0) = 31$ (the initial transmission attempt), $W(1) = 63$ (the first retransmission attempt), $W(2) = 127$ (the second retransmission attempt), $W(3) = 255$ (the third re-transmission attempt), $W(4) = 511$, $W(5) = 1023$, $W(6) = 1023$, and finally $W(7) = 1023$. Different IEEE 802.11 physical layer standards could specify different values for these parameters to optimize their operations.

In order to provide guaranteed reservation of the channel and hence uninterrupted data transmission, CSMA/CA protocol can be enhanced with Request-To-Send (RTS)/Clear-To-Send (CTS) handshake and virtual carrier sense using Network

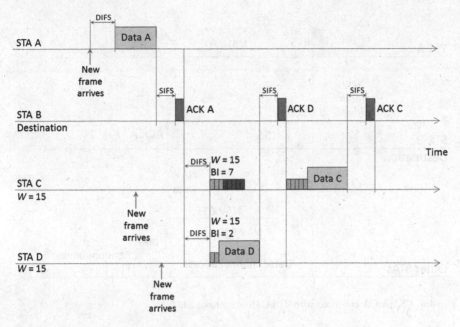

Fig. 4.4 CSMA/CA: An example of back-off procedure when there is no collision

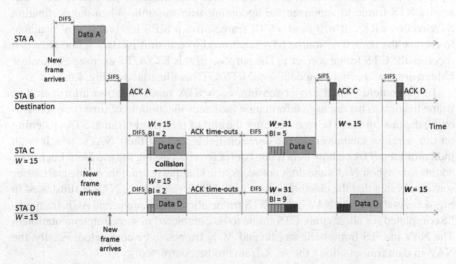

Fig. 4.5 CSMA/CA: An example of back-off procedure when there is a collision

Allocation Vector (NAV). The former is an optional mechanism and only employed for transmissions of long frames (determined by RTS threshold which is typically around 500 bytes). The latter is a prominent mechanism which is widely used with CSMA/CA protocol.

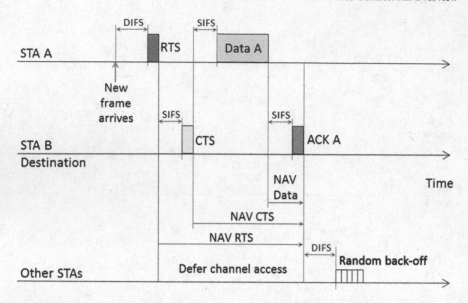

Fig. 4.6 CSMA/CA enhanced with RTS/CTS handshake and NAV

In RTS/CTS access mode, prior to the data transmission, the source STA will send a RTS frame to announce the upcoming transmission. When the destination STA receives RTS, it will send a CTS frame after a SIFS interval if it is available to receive the data. The source STA is allowed to transmit its data frame only if it receives the CTS frame correctly. The purpose of this RTS/CTS exchange is to clear hidden areas and avoid long collisions. RTS/CTS is illustrated in Fig. 4.6.

To implement virtual carrier sensing, each STA sends duration information in frame headers. This duration information indicates the amount of time (in microseconds) the medium is to be reserved after the end of the current frame. STAs listening on the wireless medium read the duration fields and set their NAVs, which is an indicator for a STA on how long it must defer from accessing the medium. Each STA counts down their NAVs and does not access the channel (even if their physical carrier sense indicates that the channel is free) until NAVs reach zero. NAV is illustrated in Fig. 4.6. As shown, the NAV field in RTS frame allows CTS, data, and ACK frames to be completed (or allows only CTS frame to be completed in some implementations). The NAV in CTS frames allows data and ACK frames to be completed. Finally, the NAV in data frames allows the ACK frame to be completed.

4.3   IEEE 802.11e EDCA

The biggest limitation of IEEE 802.11 CSMA/CA is its lack of capabilities to differentiate frames in terms of channel access priorities for different applications. As a result, the IEEE developed enhancements in IEEE 802.11e to both coordination modes to facilitate QoS. The following sections will present details of 802.11 CSMA/CA protocol and its enhancements introduced in 802.11e.

The enhancement to DCF, namely enhanced distribution coordination function (EDCF), introduces the concept of access categories (ACs) [3]. Each STA has four kinds of ACs that define four respective priority levels to differentiate the channel access probability for different traffic types. With EDCF, high priority traffic has a higher chance of being sent than low priority traffic: A STA with high priority traffic waits a little less before it sends its packet, on average, than a STA with low priority traffic. This is accomplished by using a shorter contention window and shorter arbitration interframe space (AIFS).

IEEE 802.11e extends the polling mechanism of PCF with the Hybrid Coordination Function (HCF). The HCF-controlled channel access (HCCA) works similarly to PCF. However, in contrast to PCF, in which the interval between two beacon frames is strictly divided into two periods of CFP and CP, the HCCA allows CFPs to be initiated at almost any time during a CP. This kind of CFP is called a Controlled Access Phase (CAP) in 802.11e. A CAP is initiated by the AP whenever it wants to send a frame to a STA or receive a frame from a STA in a contention-free manner. In fact, the CFP is a CAP too. During a CAP, the Hybrid Coordinator (HC), which is also the AP, controls the access to the medium using polling mechanism. During the CP, all STAs function in EDCA. The second difference with PCF is that Traffic Class (TC) and Traffic Streams (TSs) are defined. This means that HC is not limited to per-station queuing and can provide a kind of per-session service. Also, HC can coordinate these streams or sessions in any fashion it chooses (not just round robin). Moreover, STAs give information about the lengths of their queues for each TC. HC can use this information to give priority to one STA over another, or better adjust its scheduling mechanism.

IEEE 802.11e additionally introduces the concept of transmission opportunity (TXOP). A STA which obtains medium access must not utilize radio resource for duration longer than a limit specified by TXOP. The use of TXOPs reduces the problem of low-rate STAs gaining an inordinate amount of channel time in the conventional 802.11 DCF MAC. Another enhancement is that a STA is only allowed to initiate a frame exchange if it can complete the exchange before the start of the next beacon interval.

4.3.1   EDCA and HCCA

Basic operations of HCCA are illustrated in Fig. 4.7 [3]. HCCA is generally considered as the most advanced and complicated coordination function. With HCCA, QoS can be configured with great precision. QoS-enabled STAs have the ability to request specific transmission parameters (data rate, jitter, etc.), which should allow advanced applications such as voice over IP (VoIP) and video streaming to work more effectively on Wi-Fi networks. However, due to its complexity and signaling overhead, HCCA has not been widely implemented.

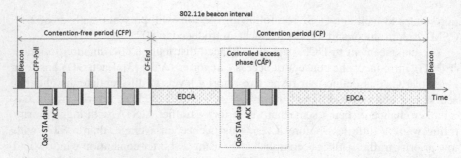

Fig. 4.7   HCCA in IEEE 802.11e

It can be seen that IEEE 802.11 CSMA/CA is the most fundamental protocol for medium access in WLANs. In fact, IEEE 802.11e EDCA is primarily designed based on CSMA/CA. As a result, in-depth knowledge on medium access mechanisms employed by this protocol is imperative to understand the potential coexistence issues which may arise.

4.4   Important Observations on CSMA/CA

It is important to note that IEEE 802.11 CSMA/CA is specified with a few key additional features that go beyond LBT requirements specified by ETSI [2, 3]. A Wi-Fi device defers to signals that are much weaker than the minimum level required by ETSI. ETSI LBT requires a transmitter to defer if the received energy is above −60 dBm (for 20 MHz), while Wi-Fi defers if the received energy is above −62 dBm (this level is referred to as the energy detect threshold, or ED for short) or if a valid Wi-Fi preamble is detected. Wi-Fi's ED threshold is nearly the same as ETSI's LBT threshold, but Wi-Fi preamble detection is required to work to at least −82 dBm, and in reality works to −90 dBm or lower in most products. Hence, Wi-Fi devices defer to other Wi-Fi transmissions much more conservatively (i.e., at a much larger distance) than a device which only meets ETSI requirements. Second, Wi-Fi goes beyond the ETSI requirements in specifying how long a device must wait after the on-air energy falls below the threshold before initiating a transmission. Finally, when a collision is inferred from the loss of a transmission, Wi-Fi employs an exponential back-off rule that doubles the contention window size and thus significantly increases the random back-off time in order to avoid future collision.

References

1. J. Berg, "The IEEE 802.11 standardization Its history, specifications, implementations, and future"; TechnicalReport GMU-TCOM-TR-8, George Mason University. Available: http://telecom.gmu.edu/sites/default/files/publications/Berg_802.11_GMU-TCOM-TR-8.pdf.
2. *ETSI EN 301 893 V1.7.2 (2014-07): Broadband radio access networks (BRAN); 5 GHz high performance RLAN; Harmonized EN covering the essential requirements of article 3.2 of the R&TTE Directive*, European Telecommunications Standards Institute Std., 2014.
3. "IEEE Standard for information technology–Telecommunications and information exchange between systems Local and metropolitan area networks–Specific requirements Part 11: Wireless LAN medium access control (MAC) and Physical Layer (PHY) specifications," *IEEE Std 802.11-2012 (Revision of IEEE Std 802.11-2007)*, March 2012.
4. "IEEE Standard for information technology–Telecommunications and information exchange between systems Local and metropolitan area networks–Specific requirements Part 11: Wireless LAN medium access control (MAC) and Physical Layer (PHY) specifications–Amendment 4: Enhancements for very high throughput for operation in bands below 6 GHz," *IEEE Std 802.11ac-2013 (Revision of IEEE Std 802.11-2007)*, December 2013.
5. "IEEE Standard for information technology–Telecommunications and information exchange between systems Local and metropolitan area networks–Specific requirements Part 11: Wireless LAN medium access control (MAC) and Physical Layer (PHY) specifications amendment 3: Enhancements for very high throughput in the 60 GHz band," *IEEE Std 802.11ad-2012*, December 2012.
6. "Wi-Fi device shipments to surpass 15 billion by end of 2016", Online, Wi-Fi Alliance. Available: http://www.wi-fi.org/news-events/newsroom/wi-fi-device-shipments-to-surpass-15-billion-by-end-of-2016.

Chapter 5
U-LTE and Wi-Fi Coexistence: A Survey

Coexistence between U-LTE and Wi-Fi networks is a deciding factor on the acceptance of U-LTE. As a result, a large number of studies have been carried out to identify what could be the effects that U-LTE may cause to Wi-Fi and which mechanisms could be used to ensure that these two technologies share the 5 GHz unlicensed frequency band in an efficient and fair manner. This chapter presents a survey on related work to answer the following questions: (i) What issues arise from simultaneous operation of LTE and Wi-Fi in the same spectrum bands? (ii) Which technology is affected the most? and (iii) Which factors determine the impacts of U-LTE to Wi-Fi?. It also identifies the strengths and weaknesses of existing solutions and suggests potential strategies to improve performance of these two technologies.

5.1 Impacts of U-LTE on Wi-Fi Operation and Performance

In [6], extensive simulations were performed to assess the performance of LTE and Wi-Fi when the technologies are coexisting in an office environment. Both single-floor and multi-floor office environments, with different assumptions as to the density of Wi-Fi and LTE nodes, have been considered. The simulation results show that, in the absence of any modification to the LTE channel access mechanism, channel sharing between LTE and Wi-Fi networks is significantly unfair for Wi-Fi networks. While LTE nodes experience only marginal performance loses when Wi-Fi is present on the same band (about 4 % from the baseline performance), in a sparse deployment of 1 AP per system per floor, Wi-Fi could lose up to 70 % compared to its baseline performance. In a dense deployment of 5 APs per system per floor, the losses seen by Wi-Fi were nearly 100 %. Detailed investigations into [6] indicate that for Wi-Fi the channel is blocked when LTE interference is present, and thus, Wi-Fi nodes remain in "listen" for a clear channel mode most of the time.

The authors in [11] present observations similar to those in [6] on the effects of unmodified LTE on the performance of Wi-Fi networks sharing the same frequency

© The Author(s) 2017   43
Q.-D. Ho et al., *Long Term Evolution in Unlicensed Bands*,
SpringerBriefs in Electrical and Computer Engineering,
DOI 10.1007/978-3-319-47346-8_5

band. Specifically, it was found that as network load is increased LTE performance suffers only a minor degradation while Wi-Fi performance drops significantly. These results can be explained by the increasing LTE occupancy in the shared band because LTE does not follow the same rules as Wi-Fi in shared medium access. When there is ongoing transmission on the channel, Wi-Fi politely defers its transmissions while LTE will always choose to transmit and simply select a more robust transmission mode by adapting its modulation and channel-coding scheme in order to cope with the higher levels of interference. This aggressive behavior quickly results in a situation where LTE terminals take all transmission opportunities while Wi-Fi devices are locked in defer and back-off procedures. Fortunately, the results in [11] have also demonstrated that the severity of this negative impact on Wi-Fi can be efficiently controlled by restricting LTE activity.

The authors in [3] analyze the performance degradation of Wi-Fi in the presence of LTE-U with the probability of Wi-Fi accessing the channel is used as the main metric. Numerical results indicate that Wi-Fi is negatively affected by conventional LTE operation due to LTE's almost continuous transmissions subsequently blocking Wi-Fi access. Specifically, given the two modes of operations currently proposed for LTE-U in the unlicensed spectrum, the "off" period presented by the LTE protocol is too short for Wi-Fi users to access the channel. As a result, Wi-Fi is at risk of spending a significant amount of time in the "listening" mode when LTE transmissions are present in the same channel.

The work in [7] presents initial investigations on the coexistence of two versions of license-anchored U-LTE (i.e., LTE-U and LAA-LTE) and Wi-Fi in 5 GHz frequency band. Results show that LTE-U poorly coexists with Wi-Fi primarily due to two factors: (i) the incompatibility of LTE-U's duty-cycling mechanism with Wi-Fi equipment and (ii) the lack of an effective coexistence mechanism in scenarios where LTE-U and Wi-Fi devices hear each other at moderate but non-negligible power levels. Additionally, LAA-LTE with LBT does not by itself guarantee successful fair coexistence with Wi-Fi. The results in [7] were submitted to the FCC in June 2015 to demonstrate that, although any wireless technology should have the ability to utilize unlicensed spectrum within the FCC's rules, U-LTE has the potential to crowd out unlicensed services.

An experiment-based study on the effect of LTE-U to Wi-Fi is presented in [13]. The LTE signal level was set higher than the Wi-Fi clients' LBT energy detection threshold (i.e., when LTE is on, the Wi-Fi client should sense their presence and not transmit). Wi-Fi throughput and latency were measured when data are transmitted through the Wi-Fi network with varying duty cycles and periods of LTE signals. The results in [13] indicate that, as expected, increasing the LTE-U duty cycle degrades both Wi-Fi throughput and latency performance since it decreases Wi-Fi transmission opportunities accordingly. If the duty cycle period is too high, Wi-Fi latency is negatively impacted (while Wi-Fi throughput is nearly unchanged, given the same duty cycle) since Wi-Fi frames have to be buffered during long LTE "on" period. However, if the duty cycle period is configured as too low (e.g., 10 ms), Wi-Fi throughput degrades due to the fact that LTE "on" and "off" periods are too short for Wi-Fi users to both access the channel and complete their transmissions. Furthermore, the authors in [13] indicate that LTE-U duty cycle cannot strictly results in corresponding air time and throughput sharings. For example, with a duty cycle of 50%, LTE-U is likely to capture more than 50% of the channel resources. The reason is that when LTE-U starts its transmissions (regardless of ongoing Wi-Fi frame transmissions), many Wi-Fi frames are corrupted. The resulting transmission failures lead to multiple frame retransmissions and, more importantly, mistakenly force Wi-Fi transceivers to operate at lower rates (in this case, lowering the channel coding and modulation modes is not necessary and is a waste of channel efficiency).

5.2 Existing Solutions to Address U-LTE and Wi-Fi Coexistence Concerns

Various coexistence mechanisms proposed for U-LTE are surveyed in [1, 2, 17]. The proposed mechanisms include: dynamic channel selection (DCS), transmission power control, opportunistic secondary cell "off," carrier sense adaptive transmission (CSAT) (in LTE-U), and LBT (in LAA-LTE). When U-LTE and Wi-Fi share the common 5 GHz radio frequency band, those mechanisms are found to be useful to reduce RFI and improve the spectrum utilization efficiency. However, the roles and performance of each varies greatly depending on network and system parameters including network scale, node density, deployment and radio environment (i.e., indoor, outdoor, short range, and long range), and network load profiles.

In order to see how LBT mechanisms employed by LAA-LTE can help promote fair coexistence, a simulation-based study is carried out and reported in [4]. LBE LBT specified by ETSI and IEEE 802.11e Enhanced Distributed Channel Access (EDCA) are assumed for LAA-LTE and Wi-Fi, respectively. The most important observation from [4] is that LBT compliant to ETSI regulation is not sufficient for fair coexistence: Wi-Fi STAs have much lower probability of successful channel access compared to LAA-LTE users. One major reason for this phenomenon is the non-exponential back-off employed by LAA-LTE LBT. Unfortunately, no form of exponential back-off LBT is studied in [4].

In [9, 10, 12, 14], the performance of LTE-U and LAA-LTE and Wi-Fi in a shared frequency band is evaluated. DCS and opportunistic secondary cell "off" in unlicensed spectrum (U-LTE small cells would release the unlicensed carriers and fall back to the anchor carrier in licensed spectrum at low traffic load) are jointly used with CSAT and LBT. The results show that coexistence has a negative but controllable impact on Wi-Fi performance. In [9, 10, 14], LTE-U can be a better neighbor to Wi-Fi than Wi-Fi would be to itself, in some scenarios. The underlying design that allows LTE-U to achieve high spectral efficiency while being a good neighbor to Wi-Fi is achieved through a set of carefully designed coexistence techniques, including DCS, secondary cell "duty cycle" in unlicensed spectrum (i.e., CSAT), and opportunistic secondary cell "off" in unlicensed spectrum. Specifically, in scenarios where the density of Wi-Fi APs and small cells is low or moderate, DCS and opportunistic secondary cell "off" are sufficient to meet the coexistence requirement. When LTE-U devices replace Wi-Fi devices, they can achieve significantly higher throughputs due to their high spectral efficiency. In addition, the performance of neighboring Wi-Fi is unchanged or even slightly improved since LTE-U devices can finish transmissions faster and incur less interference, the similarly deployed Wi-Fi APs. However, if the density of Wi-Fi devices and LTE-U small cells is high, DCS and opportunistic secondary cell "off" alone cannot guarantee harmonious coexistence with Wi-Fi and therefore, CSAT or LBT is required. Results in [10, 12] were submitted to the FCC in 2015 to support U-LTE technologies.

A systematic and large-scale network-wide study of LAA-LTE and Wi-Fi performance in a wide range of realistic deployment scenarios and network densities in the unlicensed 5 GHz band is presented in [16]. The simulation results in all considered coexistence scenarios demonstrate that both LAA-LTE and Wi-Fi significantly benefit from the large number of available channels and the isolation provided by building shielding at 5 GHz. They also suggest that deploying LAA-LTE with a random channel selection scheme is feasible for lower network densities. For typical indoor deployments of high density, implementing LTE-U interference-aware channel selection with respect to Wi-Fi is superior to LBT in terms of achieved throughput for both technologies. Additionally, LBT can increase LAA-LTE user throughput when multiple outdoor LAA-LTE networks deployed by different cellular operators coexist.

The work in [8] investigates the behavior and performance of two existing LBT mechanisms that are designed following the coexistence standard specified by ETSI, namely LBE- and FBE-based mechanisms described in Chap. 2. The Jain's fairness index has been used to assess the coexistence of LAA-LTE using these two LBT mechanisms and compare to Wi-Fi using CSMA-CA. The simulations in [8] show that the FBE-based mechanism, using a fixed contention window, impairs the channel access opportunity of Wi-Fi's CSMA-CA using an adaptive contention window. The simulations reveal that FBE-based mechanism tends to aggressively occupy the channel. In some cases, Wi-Fi devices are starved with very few, or even no, chances to access the channel. This poor fairness is mainly caused by the short CCA-sensing period of FBE-based mechanism. CCA is applied only once and then FBE-based mechanism may start its transmission immediately while LBE and Wi-Fi-based mechanisms are still decrementing their respective back-off counters. The fairness is worsened by the longer frames used by FBE. Another observation is that, again due to equal CCA sensing time, when multiple FBE are contending for the channel, they are prone to serious collisions (if they are accidentally synchronized) or suffer a significant unfairness (if they are asynchronous). To cope with those issues, tuning the values of the back-off scaler (q) to extend the contention window size and using CCA procedure similar to that of LBE-based mechanism have been suggested for LBE- and FBT-based mechanism, respectively. The results in [8] demonstrate that the modified LBE-based mechanism still cannot sufficiently improve the level of fairness achieved with others. This could be because simply empirically tuning the back-off scaler, while keeping the CCA principle unchanged, cannot compensate for the exponential growth of window size adopted by Wi-Fi's CSMA-CA. The modified FBE-based mechanism can offer better fairness when coexisting with Wi-Fi.

A comparison of LTE-U and LAA-LTE is presented in [5]. The analysis in [5] shows that for sufficiently long LTE transmission times, the LTE throughputs achieved by CSAT and LBE are almost identical. However, for shorter LTE transmission times, LTE-U provides lower LTE throughput than LAA-LTE due to higher LTE/Wi-Fi collision probability of LTE-U. Besides, while shorter LTE transmission time decreases the tail of the Wi-Fi delay distribution, the percentage of packets that suffer from long delays increases. The results also indicate that when appropriately configured, LTE-U and LAA-LTE provide the same level of fairness to Wi-Fi. The selection of coexistence mechanisms is primarily driven by the operator's interests that include implementation complexity, LTE throughput, and operational and management costs as well as strategic decisions on targeted markets.

Coordinated coexistence between U-LTE and Wi-Fi is investigated in [2, 15]. The authors in [2] propose a method of centralized system management which combines LTE-U and Wi-Fi through network function virtualization (NFV) interconnections. This technique may enable seamless transfers of resources between LTE-U and Wi-Fi using in-the-cloud control of distributed APs. However, only *conceptual* network architectures and mechanisms are presented. The authors in [15] present a software-defined networking (SDN) architecture to support logically centralized dynamic spectrum management involving multiple autonomous networks to improve spectrum utilization and facilitate coexistence. The basic design goal is to support the seamless communication and information dissemination required for coordination of heterogeneous networks. The system consists of two-tiered controllers mainly responsible for the control plane. The Global Controller (GC) acquires and processes global network state information (radio coverage maps, coordination algorithms, policy and network evaluation matrices, etc.) and controls the flow of information between Regional Controllers (RCs) and databases based on authentication and other regulatory policies. RCs acquire local visibility needed for radio resource allocation at wireless devices (device location, frequency band, duty cycle, power level, and data rate). Joint transmission power control and time division channel access optimizations are further proposed. Analytical results in [15] demonstrate that, *with full buffer traffic assumption*, centralized optimization approaches can provide fair access to the spectrum for LTE-U and Wi-Fi networks.

An experimental evaluation of U-LTE interference effects on Wi-Fi performance under various network conditions, along with some suggestions for better coexistence of U-LTE and Wi-Fi networks, is presented in [7]. Various system parameters (bandwidth, center frequency, etc.) are swept to identify the most significant parameters that determine the levels of LTE interference introduced to Wi-Fi carrier sense and their effects on performance. The results indicate that Wi-Fi throughput can be heavily degraded by LAA-LTE transmissions with 3/5/10 MHz bandwidth (especially 3/5 MHz). Besides, LAA-LTE transmissions can have small impact on Wi-Fi throughput when using a 1.4 MHz channel with center frequencies located on the guard bands or the center frequencies of Wi-Fi channels. However, the authors in [7] do not clearly define what LAA-LTE really mean in their work. It seems to be that they simply perform experiments with conventional LTE transceivers of varying power spectral densities and do not incorporate any coexistence mechanism into the LTE system.

References

1. F. Abinader, E. Almeida, F. Chaves, A. Cavalcante, R. Vieira, R. Paiva, A. Sobrinho, S. Choud-hury, E. Tuomaala, K. Doppler, and V. Sousa, "Enabling the coexistence of LTE and Wi-Fi in unlicensed bands," *IEEE Communications Magazine*, vol. 52, no. 11, pp. 54–61, Nov 2014.
2. A. Al-Dulaimi, S. Al-Rubaye, Q. Ni, and E. Sousa, "5G communications race: Pursuit of more capacity triggers LTE in unlicensed band," *IEEE Vehicular Technology Magazine*, vol. 10, no. 1, pp. 43–51, March 2015.
3. A. Babaei, J. Andreoli-Fang, and B. Hamzeh, "On the impact of LTE-U on Wi-Fi performance," in *2014 IEEE 25th Annual International Symposium on Personal, Indoor, and Mobile Radio Communication (PIMRC)*, Sept 2014, pp.1621–1625.
4. J. P. A. Babaei and J. Andreoli-fang, "Overview of EU LBT and its effectiveness for coexistence of LAA LTE and Wi-Fi," IEEE 802.19-14/0082r0, CableLabs, Nov. 2014.
5. C. Cano and D. J. Leith, "Unlicensed LTE/WiFi coexistence: Is LBT inherently fairer than CSAT?" *Scientific Publication Data: Computer Science - Networking and Internet Architecture*, nov 2015.
6. A. M. Cavalcante *et al.*, "Performance evaluation of LTE and Wi-Fi coexistence in unlicensed bands," in *Proceedings of 2013 IEEE Vehicular Technology Conference (VTC Spring)*, 2013, pp. 1–6.
7. Y. Jian, C.-F. Shih, B. Krishnaswamy, and R. Sivakumar, "Coexistence of Wi-Fi and LAA-LTE: Experimental evaluation, analysis and insights," in *2015 IEEE International Conference on Communication Workshop (ICCW)*, June 2015, pp. 2325–2331.
8. N. Jindal and D. Breslin, "LTE and Wi-Fi in unlicensed spectrum: A coexistence study," white paper, Google.
9. A. Kanyeshuli, "LTE in unlicensed band: Medium access and performance evaluation," Master's thesis, University of Agder, Norway, May 2015.
10. "LTE in unlicensed spectrum: Harmonious coexistence with Wi-Fi," whitepaper, Qualcomm Inc., Jun. 2014.
11. "LTE-U technical report: Coexistence study for LTE-U SDLV1.0(2015-02)," LTE-U Forum, Tech. Rep., 2015.
12. T. Nihtila, V. Tykhomyrov, O. Alanen, M. Uusitalo, A. Sorri, M. Moisio, S. Iraji, R. Rata-suk, and N. Mangalvedhe, "System performance of LTE and IEEE 802.11 coexisting on a shared frequency band," in *2013 IEEE Wireless Communications and Networking Conference (WCNC)*, April 2013, pp. 1038–1043.
13. "Office of engineering and technology and wireless telecommunications bureau seek informa-tion on current trends in LTE-U and LAA technology," Comments of Qualcomm Incoporated, Qualcomm Inc., Jun. 2015.
14. J. Padden, "Wi-Fi vs. duty cycled LTE: A balancing act," CableLabs. [Online]. Available: http://www.cablelabs.com/wi-fi-vs-duty-cycled-lte/
15. A. Sadek, T. Kadous, K. Tang, H. Lee, and M. Fan, "Extending LTE to unlicensed band - merit and coexistence," in *Communication Workshop (ICCW), 2015 IEEE International Conference on*, June 2015, pp. 2344–2349.
16. S. Sagari, S. Baysting, D. Saha, I. Seskar, W. Trappe, and D. Raychaudhuri, "Coordinated dynamic spectrum management of LTE-U and Wi-Fi networks," in *2015 IEEE International Symposium on Dynamic Spectrum Access Networks (DySPAN)*, Sept 2015, pp. 209–220.
17. A. Voicu, L. Simic, and M. Petrova, "Coexistence of pico- and femto-cellular LTE-unlicensed with legacy indoor Wi-Fi deployments," in *2015 IEEE International Conference on Commu-nication Workshop (ICCW)*, June 2015, pp. 2294–2300.
18. H. Zhang, X. Chu, W. Guo, and S. Wang, "Coexistence of Wi-Fi and heterogeneous small cell networks sharing unlicensed spectrum," *IEEE Communications Magazine*, vol. 53, no. 3, pp. 158–164, March 2015.

Chapter 6
Network-Aware Adaptive LBT (NALT) Coexistence Mechanism

In the absence of coordination between radio access technologies (RATs), and with the goal of deploying unlicensed LTE without requiring changes to the Wi-Fi MAC layer, it falls to the LTE base stations to ensure fair coexistence. The multiple access method used in Wi-Fi is designed for fair sharing of the channel with devices operating toward the same goal. Following this paradigm, if LAA-LTE is not carefully designed to ensure fairness, it can easily lead to Wi-Fi stations being barred from the channel. The greatest gains in fair coexistence are achieved when LAA-LTE behaves in as Wi-Fi-like a manner as possible; however, this may not allow LTE to make the best use of the channel. In this chapter, a Network-aware Adaptive LBT mechanism (NALT) is presented which passively monitors both channel conditions and usage activity to maximize transmission opportunities while respecting fair sharing of the channel, all in a way that is transparent to incumbent Wi-Fi devices. Simulation results are presented demonstrating the effectiveness of NALT in providing proportional fair sharing among LAA-LTE and Wi-Fi devices.

6.1 Background and Theoretical Basis

As discussed in Chap. 4, Wi-Fi employs a fairly simple multiple access strategy which can be easily overwhelmed if competing devices are not also designed for fair coexistence. The Wi-Fi MAC protocol employs Listen-Before-Talk (LBT) and is based on a probabilistic model of channel access which minimizes collisions through the use of random backoff to limit the probability that two stations will transmit at the same time after the channel has become idle [3]. When a collision is inferred after a failed transmission, the set of possible backoff values grows exponentially to further reduce the probability of subsequent transmission failures. While the ETSI LBT standard, on which the recommended mechanism for LAA-LTE is based, is also probabilistic, it employs a random backoff from a fixed set of possible backoff values [1], which does not attempt to reduce the probability of collision on repeated failed transmission. Thus, if a collision occurs, Wi-Fi will react by reducing its

Q.-D. Ho et al., *Long Term Evolution in Unlicensed Bands*,
SpringerBriefs in Electrical and Computer Engineering,
DOI 10.1007/978-3-319-47346-8_6

probability of gaining access to the channel, while a device modeled on the ETSI LBT mechanism will maintain the same probability of channel access. Additionally, LAA-LTE used for supplemental downlink or carrier aggregation is expected to align subframes with the licensed band, and such subframes have a duration of 1ms, which can be significantly longer than the average channel occupancy time of a Wi-Fi station. Combined, these two factors will lead to LAA-LTE stations both winning the channel more frequently and then occupying the channel for significantly longer than an average competing Wi-Fi station would, even if the number of channel accesses was equal. Since Wi-Fi stations may operate at any of several modulation and coding schemes, it is also difficult to provide throughput fairness across a large number of Wi-Fi devices. However, airtime fairness can be achieved by leveraging the principles developed for the 802.11e Enhanced Distributed Channel Access (EDCA) function for service differentiation between traffic priorities in Wi-Fi.

In EDCA, Wi-Fi parameters such as contention window and inter-frame spacing are set up to provide quality-of-service differentiation and priority enforcement between varying types of traffic [3]. By changing these parameters, it is possible to impact the probability of channel access in a predictable way. By constantly managing these parameters in response to network activity, it is possible to maintain long run proportionally fair sharing between the traffic categories or classes of devices on competing networks.

Specifically, the relationship between minimum contention window size for two traffic classes and their relative proportion of channel access has been found to be

$$\frac{\theta_i}{\theta_j} \approx \frac{CW_{min}^j}{CW_{min}^i} \tag{6.1}$$

where $\frac{\theta_i}{\theta_j}$ is the ratio of channel access class i receives to that received by class j, and CW_{min}^x is the minimum contention window used by class x [2, 5].

To use Eq. 6.1 to balance airtime between LAA-LTE and Wi-Fi, it is necessary to treat all Wi-Fi stations and all LAA-LTE networks as traffic classes and account for the duration of channel access for each class. Between the two traffic classes, this duration will generally be longer for LAA-LTE than for Wi-Fi, due to the synchronization between licensed and unlicensed transmissions and the range of data rates available for Wi-Fi stations over clean channels. As such, LAA-LTE will receive fewer channel accesses in order to achieve the same airtime allocation. For example, if a Wi-Fi channel access takes half of the time of a LAA-LTE channel access, in the case of a single LAA-LTE station competing with a single Wi-Fi station, then the Wi-Fi station should receive twice as many transmission opportunities as the LAA-LTE station in order to achieve equal airtime. If there were two Wi-Fi stations, in order for each to have equal airtime, the LAA-LTE station should receive one-quarter as many transmission opportunities as the combined Wi-Fi stations, so that proportionally each of the three stations would receive equal airtime on average. Adding a proportionality constant ρ, which is the ratio of LAA-LTE transmission time to average Wi-Fi transmission time, and solving Eq. 6.1 for the required CW_{min}

values to realize equal airtime yield

$$CW_{min}^{LTE} = \rho \cdot CW_{min}^{WiFi} \tag{6.2}$$

Since we seek equal airtime, we require that $\rho \cdot \frac{\theta_{WiFi}}{\theta_{LTE}} = 1$, or in other words, the Wi-Fi traffic class receives ρ times as many channel accesses as the LAA-LTE class.

The relation in Eq. 6.2 provides an approximation of the optimal CW_{min}^{LTE} to provide airtime fairness; however, the Wi-Fi traffic class may be made up of stations which are using different transmission rates and CW_{min} values. In order to estimate the CW_{min}^{WiFi} to use in Eq. 6.2, and adjust to changing network topologies, an estimate of the average current Wi-Fi contention window being used is required. Such an estimate can be obtained from the relationship between contention window and the probability of collision. For Wi-Fi networks, the probability of collision, p, in a saturated network is given by,

$$p = 1 - (1 - 1/CW_{avg})^{n-1} \tag{6.3}$$

where CW_{avg} is the average contention window currently being employed in the network, and n is the number of competing stations [4]. Rearranging and solving for CW_{avg} yield

$$CW_{avg} = \frac{1}{1 - e^{ln(1-p)/(n-1)}} \tag{6.4}$$

Equation 6.4 provides the average contention window size for all stations, both LAA-LTE and Wi-Fi, i.e., $n = n_{WiFi} + n_{LTE}$. In order to consider only the average contention window size for the Wi-Fi stations, and noting the optimal CW_{min}^{LTE} to CW_{min}^{WiFi} ratio, we can estimate CW_{min}^{WiFi} as

$$CW_{avg}^{WiFi} = CW_{avg} \left(\frac{n_{WiFi} + n_{LTE}}{n_{WiFi} + \rho \cdot n_{LTE}} \right) \tag{6.5}$$

Combining Eq. 6.2 through Eq. 6.5, we set

$$CW^{LTE} = \rho \cdot CW_{avg}^{WiFi} = \frac{\rho}{1 - e^{ln(1-p)/(n_{WiFi}+n_{LTE}-1)}} \left(\frac{n_{WiFi} + n_{LTE}}{n_{WiFi} + \rho \cdot n_{LTE}} \right) \tag{6.6}$$

Adapting the contention window used in each LAA-LTE network according to Eq. 6.6 will provide proportional fair channel access across the two classes of devices in the long run. In order to make use of this relation, the LAA-LTE base stations must know, or be reasonably able to estimate, the probability of collision in the network, p, as well as the number and type of competing devices.

6.2   Proposed Mechanism

NALT is defined as a simple, distributed coordination function to be implemented by LAA-LTE base stations. This allows several LAA-LTE networks to effectively and independently fairly share the channel with both each other and incumbent Wi-Fi stations, without any changes being required in the Wi-Fi stations.

To make use of the relationships in the previous sections, the following assumptions are made:

- NALT-enabled base stations are able to:
  - Analyze traffic on the channel and determine the number of competing stations and their types
  - Determine average transmission durations either by decoding transmission headers, actively timing the transmissions, or some other suitable mechanism
- Successfully gaining access to the channel means that the transmission was successful, i.e., ignoring noise sources and the hidden terminal problem, which NALT does not attempt to address
- Failed LAA-LTE transmissions on unlicensed channels can be reported to the base station on control channels in the licensed spectrum
- Collisions experienced on the LAA-LTE network occur with approximately the same probability as collisions experienced by Wi-Fi stations

In order to use Eq. 6.6 in an implementable algorithm, the probability of collision p and the number of competing Wi-Fi stations and LAA-LTE networks are required. Since these values cannot be known beforehand, the number of competitors is learned over time and the required probability of collision is estimated in each NALT-enabled LAA-LTE network as the ratio of observed LAA-LTE collisions to the number of LAA-LTE channel uses, on a network by network basis. Noting that this is an empirical estimate of the true statistic, its reliability is inversely proportional to the number of samples and, although it improves over time, it must be considered highly suspect for a limited number of samples and be restricted to some reasonable range.

The relationship in Eq. 6.6 is exploited to achieve fair airtime allocation by tuning the CW_{min} values used by competing stations in each LAA-LTE network. Since it is desired to avoid any changes to Wi-Fi, and fairer coexistence can be achieved by designing a more "Wi-Fi-like" MAC layer for LAA-LTE, the contention window used by LAA-LTE must increase as the number of collisions increases. To facilitate fair airtime allocations across all competing devices, the contention window should follow Eq. 6.6. Based on the limitations of the estimates employed, and to ensure that the contention window stays within reasonable bounds, the maximum and minimum values for CW^{LTE} are chosen to match the range of possible values for Wi-Fi [3].

Combining these requirements and the preceding equations and assumptions, at each time instance an LAA-LTE station will estimate the average Wi-Fi contention window as follows:

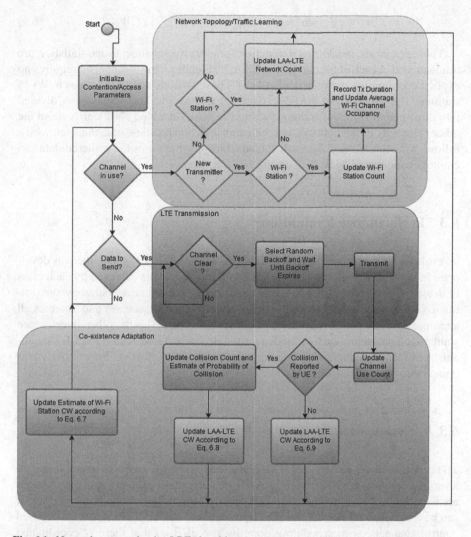

Fig. 6.1 Network-aware adaptive LBT algorithm

$$CW_{avg}^{WiFi} = CW_{min}^{WiFi}, \text{ if } \{ \# \text{ of Wi-Fi Tx} \} > \rho \cdot \{ \# \text{ of LAA-LTE Tx} \}$$

$$\text{Otherwise, update according to Eq. 6.5.} \quad (6.7)$$

LAA-LTE will follow the same backoff procedure as Wi-Fi by increasing its contention window after a collision according to

$$CW^{LTE} = min \left[max \left(CW^{LTE} * 2, \rho \cdot CW_{avg}^{WiFi} \right), CW_{MAX}^{LTE} \right] \quad (6.8)$$

and decreasing its contention window after a successful transmission according to

$$CW^{LTE} = \ min\left[max\left(CW_{MIN}^{LTE}, \rho \cdot CW_{avg}^{WiFi}\right), CW_{MAX}^{LTE}\right] \qquad (6.9)$$

These equations, in addition to function for gathering channel usage statistics, are implemented in each of the competing LAA-LTE eNBs. This mechanism requires no explicit coordination between the LAA-LTE base stations nor any changes to Wi-Fi stations. The operation of NALT is depicted in Fig. 6.1. The operation is divided into three main functions: learning, where the NALT-enabled eNB learns about the other occupants of the network and their transmission profiles; transmission, which follows a random backoff interval; and adaptation, which uses the gathered data and employs exponential backoff to ensure fair coexistence.

6.3   Performance Evaluation

To evaluate the performance of NALT, a high-level MATLAB simulation was developed in which the proportion of successful channel accesses achieved by each class of devices was tracked. Since NALT is an adaptive medium access strategy, the simulation looks only at the proportion of successful channel accesses and assumes all attempted transmissions only fail if a collision occurs. That is, other interference sources and problems, such as hidden terminals, are ignored. The ETSI LBT mechanism was simulated as a benchmark against which the effectiveness of NALT was compared.

6.3.1   System Model

The system includes a number of NALT-enabled LAA-LTE networks interacting with a varying number of Wi-Fi devices. LAA-LTE transmissions in the unlicensed bands are expected to be aligned with the LTE-A frames in the licensed spectrum; thus, it can be assumed that LAA-LTE user equipment (UE) will be coordinated via licensed control channels, with scheduling done by the eNB so that there is coordinated channel accesses for both uplink and downlink traffic. The system model incorporates this by assuming the eNB will not schedule either its own DL transmission or two UE UL transmissions in the same time–frequency slot, so that the only sources of collisions are from Wi-Fi stations and other LAA-LTE networks. Thus, each simulated LAA-LTE device in fact represents an independent network of LAA-LTE devices which are not required to contend with each other. Additionally, although the NALT-enabled eNB would be capable of analyzing traffic on the channel to determine the average Wi-Fi transmission parameters, such as bit rate and channel occupancy time, for simplicity, we assume that both UE and Wi-Fi stations use the same modulation and coding scheme and channel bandwidth, resulting in a data rate of 135 Mbps. Other than the adaptive contention window, the LAA-LTE channel occupancy and minimum idle time were modeled after ETSI LBE LBT and the

Table 6.1 NALT simulation parameters

Parameter	Value
Number of competing Wi-Fi stations	$1 - 15$
Wi-Fi OFDM symbol duration (slot)	$9\,\mu s$
DCF inter-frame spacing[a]	$34\,\mu s$
Short inter-frame spacing[a]	$16\,\mu s$
Wi-Fi frame Size	1536 bytes
Wi-Fi Tx duration[b] (Frame Tx + SIFS +ACK)	$198\,\mu s$
Number of independent LTE networks	1, 5
LAA-LTE channel occupancy time	$1000\,\mu s$

[a]Defined inter-frame spacing per 802.11n operating in the 5 GHz band
[b]Based on header transmitted at lowest supported rate and remaining frame at specified bit rate

proposed mechanisms for LAA-LTE [1]. The other pertinent simulation parameters are listed in Table 6.1.

6.3.2 Simulation Results

NALT is a probabilistic coexistence mechanism, so to evaluate the fairness provided by NALT the average of numerous trials was considered. Network topologies of between 1 and 15 Wi-Fi stations contending with LAA-LTE networks were examined and the proportion of successful channel accesses for each class of devices, related to airtime by the class' transmission duration, was tracked across all trials.

Initially, NALT was tested with a single LAA-LTE network competing with between 1 and 15 Wi-Fi stations. The resulting proportion of airtime for each device when using NALT is shown in Fig. 6.2. In each configuration, fair sharing was achieved, with every member of each class receiving a proportional airtime allocation.

For comparison, the simulation was run with the same parameters as in Table 6.1, but with a fixed contention window size of 16, corresponding to the midpoint of possible values under ETSI LBE LBT [1]. The resulting airtime allocations, normalized to the number of devices, are shown in Fig. 6.3. As expected, LAA-LTE transmission receives a disproportionately high airtime allocation compared to Wi-Fi. This is a result of the static contention window used in ETSI LBT providing an increasingly higher proportion of channel accesses as collisions on the channel occur and the Wi-Fi contention window grows.

Since it is likely that LAA-LTE networks will be deployed alongside other competing LAA-LTE networks, the simulation was extended to evaluate the effectiveness of NALT under these conditions. The simulation was run with 5 independent NALT-enabled LAA-LTE networks competing against each other as well as Wi-Fi stations.

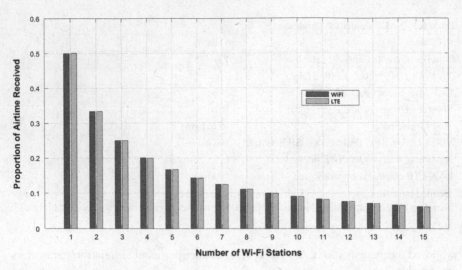

Fig. 6.2 Airtime allocations for each station with LAA-LTE using NALT

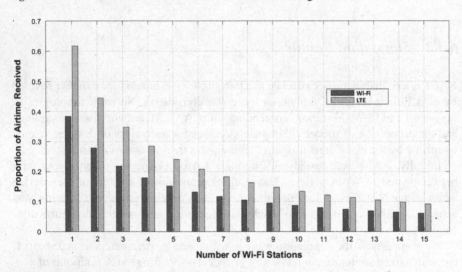

Fig. 6.3 Airtime allocations for each station with LAA-LTE using ETSI LBE LBT

The resulting proportion of airtime for each device when using NALT is shown in Fig. 6.4.

It is further conceivable that LAA-LTE networks will be deployed where there are either no competing Wi-Fi stations, or the level of interference between the RATs is negligible. Figure 6.5 shows the resulting fair allocation of airtime for each device when NALT is used in a LAA-LTE only deployment.

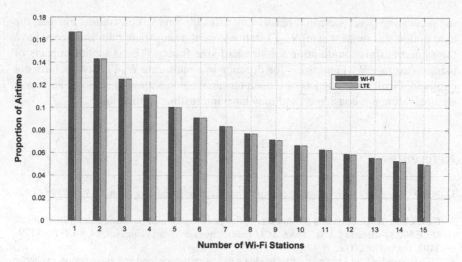

Fig. 6.4 Airtime allocations for each station with five LAA-LTE networks using NALT

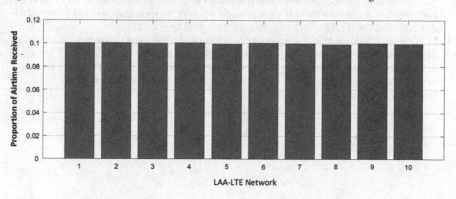

Fig. 6.5 Airtime allocations from LAA-LTE only channel contention when using NALT

6.4 Discussion and Future Work

NALT requires no changes to Wi-Fi devices, and high-level simulation results show promise in providing fair coexistence in several deployment scenarios. In each of the cases examined, NALT provides approximately equal airtime to each station, regardless of type or how many competing stations exist.

As noted, several simplifying assumptions were made, which may affect the results. It is reasonable that LAA-LTE would be able to analyze the channel and determine the number or competing Wi-Fi stations as well as their transmission profiles, from the Wi-Fi preamble and MAC header; however, a learning period to gather sufficient data to make reasonable estimates of the averages may or may not be necessary. If necessary, it may negatively impact overall performance. It is desirable to implement the learning functions depicted in Fig. 6.1 and determine whether the processing overhead could reasonably meet the timing constraints. Further, the assumption was made that all Wi-Fi stations use the same data rate, and hence have the same transmission duration for standard size frames. The potential impacts of using an average Wi-Fi transmission duration in a multi-rate Wi-Fi network have not explored. Finally, the impacts of hidden terminals, non-saturated stations, and lossy channels were not considered and may have interesting implications.

References

1. 3GPP, "Feasibility study on licensed-assisted access to unlicensed spectrum," *TR 36.889 v13.0.0*, July 2015.
2. C. T. Chou, K. G. Shin, and S. Shankar, "Contention-based airtime usage control in multirate IEEE 802.11 wireless LANs," *IEEE/ACM Transactions on Networking*, vol. 14, no. 6, pp. 1179–1192, December 2006.
3. "IEEE Standard for information technology–Telecommunications and information exchange between systems Local and metropolitan area networks–Specific requirements Part 11: Wireless LAN medium access control (MAC) and physical layer (PHY) specifications," *IEEE Std 802.11-2012 (Revision of IEEE Std 802.11-2007)*, pp. 818–972, March 2012.
4. H. L. Vu and T. Sakurai, "Collision probability in saturated IEEE 802.11 networks," in *Australian Telecommunication Networks & Applications Conference*, September 2006, pp. 1–5.
5. J. Yoon, S. Yun, et al, "Maximizing differentiated throughput in IEEE 802.11e wireless LANs," in *Proceedings of the 31st IEEE Conference on Local Computer Networks*, vol. 14, no. 6, November 2006, pp. 411–417.

Chapter 7
Open Questions and Potential Research Directions

Since U-LTE is a nascent LTE technology, the coexistence of this technology and Wi-Fi remains one of the most active research topics and working areas. Based on observations obtained from the survey and our study presented in Chaps. 5 and 6, this chapter attempts to highlight a number of open research questions and issues. Potential solutions to those issues are also identified. Primarily, the cooperation of LTE and Wi-Fi so that they could have a better understanding of each other when operating in the same area using the same radio frequency band is suggested. Such understanding can be used to take more vigilant action and help to avoid aggressive channel access that could corrupt ongoing transmissions and to design relevant protocols for a fair spectrum sharing.

7.1 LTE-U-Aware CSMA-CA and LTE-U with LBT

LTE-U mostly assumes neither coordination nor synchronization between itself and Wi-Fi system. LTE-U's "on" and "off" cycles are only known by LTE devices themselves. Vice versa, Wi-Fi control and management frames are known by Wi-Fi devices themselves. This independent operation results in various transmission issues. First, in cases when LTE-U's "on" duration is not sufficiently long while Wi-Fi exponential back-off procedure generates long back-off intervals, Wi-Fi STAs may not have a chance to utilize the channel when LTE-U is not active. Such a conservative channel access principle wastes the radio resources and results in Wi-Fi's poor performance. Second, an unfinished Wi-Fi frame transmission that was started during the LTE-U's "off" duration might be corrupted by the LTE frames once LTE switches to "on" cycle. Figure 7.1 visualizes two examples.

To mitigate these issues, inter-RAT communications between LTE and Wi-Fi could be employed to inform Wi-Fi system the LTE-U's "on" and "off" cycles. Wi-Fi system then can adapt its MAC protocol (i) to occupy the channel more opportunistically during LTE-U's "off" period (but not to increase the collision probability among Wi-Fi STAs) and (ii) to schedule frame transmissions in such a way that they

© The Author(s) 2017
Q.-D. Ho et al., *Long Term Evolution in Unlicensed Bands*,
SpringerBriefs in Electrical and Computer Engineering,
DOI 10.1007/978-3-319-47346-8_7

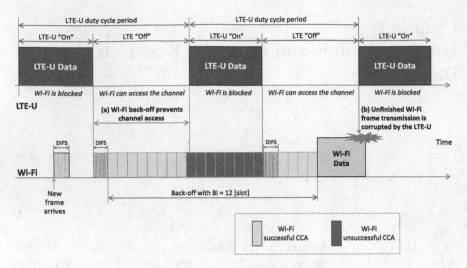

Fig. 7.1 Negative interactions between LTE-U and Wi-Fi systems

will not step on the next LTE-U's "on" cycle. Further, frame collisions could be mitigated by incorporating some form of LBT/CCA into LTE-U. Specifically, CCA should be performed before activating LTE-U's "on" cycle. If the channel is detected busy, LTE-U's "on" cycle is deferred.

7.2 LAA-LTE with Exponential Back-Off

While LBT, as a general approach, can be a good basis for coexistence of LAA-LTE and Wi-Fi, the LBE LBT in its current form (as introduced by European regulations) which is adopted for LAA-LTE is still unfair to Wi-Fi. LAA-LTE nodes impact Wi-Fi nodes in terms collision rate and probability of successful channel access more than similar Wi-Fi nodes on the same carrier. This is not compliant with the objectives as listed in 3GPP LAA-LTE Study Item [1]: "LAA should not impact Wi-Fi services (data, video and voice services) more than an additional Wi-Fi network on the same carrier; these metrics could include throughput, latency, jitter, etc." One major contributing factor is that while Wi-Fi applies exponential back-off rule, LAA-LTE simply applies fixed-size back-off rule. In order to illustrate this observation, consider the following: It is assumed that $W^{LAA} = 31$, $W^{Wi\text{-}Fi}_{min} = 31$, and $W^{Wi\text{-}Fi}_{max} = 1023$. Then, as described in Sect. 1.3.2, LTE-U always backs off with contention window $W^{LAA} = 31$. For Wi-Fi, as described in Sect. 4.2.3, it backs off with contention window $W(0) = 31$ for the initial transmission attempt. However, if collisions occur, it progressively doubles its contention windows to reduce the probability of a subsequent collision: $W(1) = 63$ (the first retransmission attempt), $W(2) = 127$ (the second retransmission attempt), $W(3) = 255$ (the third retransmission attempt),

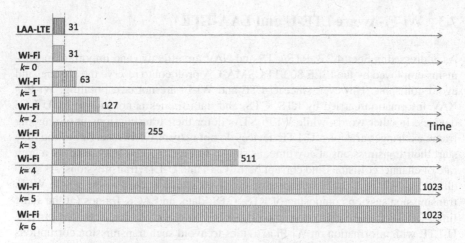

Fig. 7.2 Wi-Fi exponential back-off competes for the channel more conservatively, compared to LAA-LTE

$W(4) = 511$, $W(5) = 1023$, $W(6) = 1023$, and so on. Figure 7.2 compares how the contention windows of LAA-LTE and Wi-Fi changes in response to collisions.

At present, there is no existing work that studies how an exponential back-off can help to improve the fairness between LAA-LTE and Wi-Fi. It is important to note that, compared to Wi-Fi, designing an exponential back-off protocol for LAA-LTE that employs OFDMA-based MAC layer might not be straightforward. In detail, Wi-Fi adopts OFDM in the PHY layer and allows only one user to occupy the whole channel at one time. Its contention window is scaled, respectively, to the outcome (success or failure) of a frame transmission to given user. For LTE, OFDMA divides the system bandwidth into a series of physical resource blocks (PRBs). Each PRB is composed of 12 OFDM subcarriers. Different PRBs can be allocated to different users in a given subframe and multiple users can occupy the channel at the same time. This implies that the rule governing the adaptation of contention window of LAA-LTE is required to be more sophisticated than that of Wi-Fi. In addition to back-off procedure design, there are two other interesting questions: (i) How exponential back-off could (negatively) affect the performance and efficiency of LAA-LTE; and (ii) what could be appropriate values for LAA-LTE's operation parameters.

A side note is that, according to [3], 3GPP is now having a working agreement to use a LBT mechanism with exponential back-off. At this moment, LAA-LTE standard is not yet finalized by 3GPP and no information is publicly available. ETSI is also devising a set of minimum "fairness" requirements as part of EN 301 893 standard for "5 GHz high performance wireless access systems" in Europe (scheduled to be completed by the end of 2015).

7.3 Wi-Fi-Aware LTE-U and LAA-LTE

As addressed in Sect. 4.2.3, RTS/CTS and NAV are effective and important mechanisms employed by the IEEE 802.11 CSMA-CA protocol to reserve the channel and
avoid collisions. However, since U-LTE and Wi-Fi are not collaborating, Wi-Fi's
NAV information carried by RTS, CTS, and data frames is not known by U-LTE
devices. In other words, while Wi-Fi STAs defer their transmissions until ongoing
frame exchanges are done, U-LTE devices do not respect Wi-Fi reservation and may
start their transmissions at any time, as shown in Fig. 7.3. This may result in a high
rate of channel collisions and corrupt both Wi-Fi and U-LTE transmissions. As visu-
alized Fig. 7.3, an U-LTE transmission could accidentally destroy the whole Wi-Fi
transmission session composing of RTS, CTS, data, and ACK frames (at the same
time, U-LTE frame is also corrupted by Wi-Fi frames). Mechanisms that provide
U-LTE with information on Wi-Fi activities to avoid such transmission corruptions
could be therefore very beneficial.

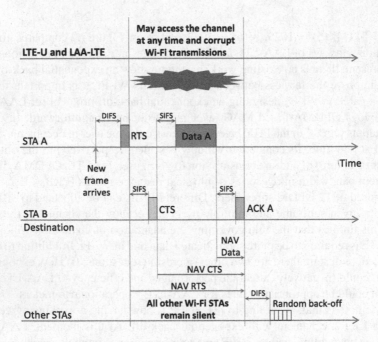

Fig. 7.3 U-LTE may cause channel collisions with Wi-Fi at any time

7.4 Collaborative U-LTE and Wi-Fi

As mentioned so far, almost all existing works dealing with U-LTE and Wi-Fi coexistence assume non-cooperative approach which does not require any information exchange between these two networks. LTE is simply additionally equipped with some mechanisms to friendly share the same channel with existing Wi-Fi networks. The authors in [2, 5] carry out preliminary investigations toward this direction. However, only conceptual network architectures and mechanisms are presented. Collaborative approaches are quite interesting since it may result in better coexistence by sharing information between different radio access technologies (RATs) and enabling global/local optimizations. Some benefits of such approaches have been outlined in Sects. 7.1 and 7.3. This approach, on the other hand, may be challenging since it needs additional network infrastructure/entities and set of protocols for inter-RAT communications. They are required for the discovery of neighboring radio systems, selecting operating channels/transmission power, etc., for radio systems, and providing some level of fair and/or efficient use of available channels.

7.5 Inter-operator U-LTE Coexistence

In addition to coexistence between U-LTE and Wi-Fi, coexistence among U-LTE systems deployed by different operators running in a shared band is also a critical concern. This concern is more pronounced in high-density urban areas with a very large number of devices/system running different protocols. Work in [6] presents a preliminary study on this and the results show that LBT mechanisms can increase the network throughput since collisions can be mitigated. Work in [4] investigates the interactions between different LBT mechanisms when they are deployed in the proximity of each other. Inter-operator U-LTE coexistence is especially important when multiple operators employ similar MAC protocols based on fixed contention windows that could be accidentally synchronized in channel access attempts and result in consecutive collisions. As a result, exponential back-off rules, inter-RAT communications, and collaborative interference management protocols could be promising approaches.

7.6 Other Considerations on Coexistence

Operations, system performance, and coexistence of radio networks highly depend on deployment scenarios. This is the main reason why a number of existing work support U-LTE technology while the others call for further investigations and developments before deploying this technology. Also, different coexistence mechanisms are recommended for different scenarios. For a complete understanding of U-LTE impacts on Wi-Fi, a wide range of node and load densities should be considered. Additionally, performance of voice- and video-related applications should be evaluated. For most of the existing work, only throughput and channel access probability of Wi-Fi networks are evaluated. However, an insight into latency and jitter performance could be desirable and it would be interesting to take into account the operations and performance of recent Wi-Fi variants when coexisting with U-LTE.

7.7 Emerging Wi-Fi Technologies and U-LTE

With the current trends of future RANs, including network densification, heterogeneous networks (HetNet), Internet of Things (IoT), and the explosion of various applications (smart homes/cities, smart transportations, autonomous vehicles, etc.), numerous technological evolutions have been emerging. For time-sensitive applications (e.g., sensor and control for critical infrastructures and autonomous vehicles), data communication is required to be extremely reliable, robust, energy-efficient while being able to guarantee latencies in millisecond or sub-millisecond scale. These requirements urge for the developments of collaborative, well-controlled, and synchronous Wi-Fi MAC protocols (instead of distributed, random access-based, and asynchronous IEEE 802.11 CSMA/CA that have been widely deployed). To this end, PCF and HCCA operation schemes (specified in IEEE 802.11/802.11e standards but not widely used) should be revisited.

Despite the fact that PCF and HCCA allocate the channel to STAs in a well-controlled manner, their performance (in terms of throughput, latency, and power consumption) is still questionable due to their complexities and signaling overhead, especially in highly dense networks with a vast number of battery-operated devices exchanging short and bursty messages. Furthermore, it is compelling to understand their interaction and coexistence with U-LTE. While CFP and CAP are desired for time-sensitive applications, the aggressive operation of U-LTE in the same frequency band may render them impossible. Finally, protocols and enabling technologies for collaborations and synchronizations between PCF-/HCCA-based Wi-Fi and U-LTE appear to be essential and thus could be very interesting working areas.

References

1. "3GPP work item: Study on licensed-assisted access using LTE to unlicensed spectrum," 3rd Generation Partnership Project (3GPP).
2. A. Al-Dulaimi, S. Al-Rubaye, Q. Ni, and E. Sousa, "5G communications race: Pursuit of more capacity triggers LTE in unlicensed band," *IEEE Vehicular Technology Magazine*, vol. 10, no. 1, pp. 43–51, March 2015.
3. M. L. Brown, "Current trends in LTE-U and LAA technology," Comments of Qualcomm Incorporated, Qualcomm Inc., Jun. 2015.

4. A. Kanyeshuli, "LTE in unlicensed band: Medium access and performance evaluation," Master's thesis, University of Agder, Norway, May 2015.
5. S. Sagari, S. Baysting, D. Saha, I. Seskar, W. Trappe, and D. Raychaudhuri, "Coordinated dynamic spectrum management of LTE-U and Wi-Fi networks," in *2015 IEEE International Symposium on Dynamic Spectrum Access Networks (DySPAN)*, Sept 2015, pp. 209–220.
6. A. Voicu, L. Simic, and M. Petrova, "Coexistence of pico- and femto-cellular LTE-unlicensed with legacy indoor Wi-Fi deployments," in *2015 IEEE International Conference on Communication Workshop (ICCW)*, June 2015, pp.2294–2300.

Printed in the United States
By Bookmasters